AIR CRASH INVESTIGATIONS

THE BOEING 737 MAX DISASTER (PART 1)

The Crash of Lion Air Flight 610

AIR CRASH INVESTIGATIONS

Over the last decades flying has become an every day event, there is nothing special about it anymore. Safety has increased tremendously, but unfortunately accidents still happen. Every accident is a source for improvement. It is therefore essential that the precise cause or probable cause of accidents is as widely known as possible. It can not only take away fear for flying but it can also make passengers aware of unusual things during a flight and so play a role in preventing accidents.

Air Crash Investigation Reports are published by official government entities and can in principle usually be down loaded from the websites of these entities. It is however not always easy, certainly not by foreign countries, to locate the report someone is looking for. Often the reports are accompanied by numerous extensive and very technical specifications and appendices and therefore not easy readable. In this series we have streamlined the reports of a number of important accidents in aviation without compromising in any way the content of the reports in order to make the issue at stake more easily accessible for a wider public.

Dirk Jan Barreveld, editor.

AIR CRASH INVESTIGATIONS

THE BOEING 737 MAX DISASTER (PART 1)

The Crash of Lion Air Flight 610

Dirk Jan Barreveld, editor.

FOUNDATION SAGIP KABAYAN

AIR CRASH INVESTIGATIONS-The Crash of Lion Air Flight 610, 29 October 2018, Komite Nasional Keselamatan Transportasi (KNKT), Transportation Building, 3rd Floor, Jalan Medan Merdeka Timur No. 5 Jakarta 10110, Indonesia, October 2019.

Lion Air Flight 610 was a scheduled domestic flight from Soekarno-Hatta Airport in Jakarta to Depati Amir Airport in Pangkai Pinang. On 29 October 2018 the Boeing 737 MAX operating the route crashed into the Java Sea 13 minutes after takeoff, killing all 189 passengers and crew.

All Rights Reserved © by Foundation Sagip Kabayan
sagipkabayan@yahoo.com

No part of this book may be reproduced or transmitted in any form or by any means, graphic, electronic, or mechanical, including photocopying, recording, taping, or by any information storage retrieval system, without the permission in writing from the publisher.

Dirk Jan Barreveld, editor
sagipkabayan@hotmail.com

A Lulu.com imprint

ISBN: 978-1-6781-6273-3

CONTENTS

7	Synopsis
9	**Chapter 1: Factual Information**
9	- History of the flight
21	- Injuries to Persons and damage to Aircraft
31	- Angle of Attack (AOA) Sensors Historical Record
43	- Autopilot System and Flight Director (F/D)
52	- Meteorological Information
52	- Aids to Navigation
54	**Chapter 2: Flight Recorders**
54	- Digital Flight Data Recorder (DFDR)
56	- Cockpit Voice Recorder
65	**Chapter 3: Additional Information**
65	- Wreckage and Impact Information
66	- Fire
66	- Survival Aspects
68	**Chapter 4: Tests and Research**
66	- Installation Test AOA Sensor with Known Bias
68	-Observation to the Xtra Aerospace
71	- Engineering Simulator
75	**Chapter 5: Organizational and Management Information**
75	- Aircraft Operator
75	- Operation Manual (OM)-part A
85	- Landing at the Nearest Suitable Airport
92	- Fault Reporting Manual (FRM)
98	- Lion Air Maintenance Management
111	- Approved Maintenance Organization
123	**Chapter 5: Federal Aviation Administration (FAA) Aircraft Certification**
123	- Directorate General of Civil Aviation in Indonesia
125	- FAA Certification Office
132	- Maneuvering Characteristics Augmentation System (MCAS) Assessment
135	- Indonesia Type Certificate Validation Process
146	**Chapter 6: Additional Information**
146	- PK-LQP Previous Flight
156	**Chapter 7: Analysis**
161	- Accident Flight Crew Actions
175	- MCAS Certification
194	**Chapter 8: Conclusions**

194	- Findings
206	- Contributing Factors
209	**Chapter 9: Safety Action**
209	- Lion Air
213	- Batam Aero Technic
214	- Boeing Company
215	- Collins Aerospace
216	- AirNav Indonesia Branch Office JATSC
217	- Federal Aviation Administration
219	- NTSB Recommendation to FAA
221	- Directorate General of Civil Aviation
222	**Chapter 10: Safety Recommendations**
222	- Lion Air
223	- Batam Aero Technic
223	- AirNav Indonesia
224	- Xtra Aerospace
224	- Boeing
226	- Directorate General of Civil Aviation (DGCA)
226	- Federal Aviation Administration (FAA)
229	**Appendices**
230	**Appendix 1: NTSB system safety and certification specialist's report**
274	**Appendix 2: Lion Air Comments**
283	**Notes**
287	**Other Air Crash Investigations**

SYNOPSIS

On 29 October 2018, at about 0632 Local Time (23:32 UTC 28 October 2018), a PT Lion Mentari Airlines (Lion Air) Boeing 737-8 (MAX) aircraft registered PK-LQP, was being operated as a scheduled passenger flight from Soekarno-Hatta International Airport (WIII), Jakarta with intended destination of Depati Amir Airport (WIPK), Pangkal Pinang, when the aircraft disappeared from radar after informing Air Traffic Controller (ATCo) that they had flight control, altitude and airspeed issues. The aircraft impacted the water in Tanjung Karawang, West Java, all person on board perished and the aircraft destroyed.

On 26 October 2018, the SPD (speed) and ALT (altimeter) flags on the Captain's primary flight display first occurred on the flight from Tianjin, China to Manado, Indonesia. Following reoccurrence of these problems, the left angle of attack (AOA) sensor was replaced in Denpasar on 28 October 2018.

The installed left AOA sensor had a 21° bias which was undetected during the installation test in Denpasar. The erroneous AOA resulted in different indications during the flight from Denpasar to Jakarta, including IAS (indicated airspeed) DISAGREE, ALT (altitude) DISAGREE, FEEL DIFF PRESS (feel differential pressure) light, activations of Maneuvering Characteristics Augmentation System (MCAS) and left control column stick shaker which were active throughout the flight. The flight crew was able to stop the repetitive MCAS activation by switched the stabilizer trim to cut out.

After landed in Jakarta, the flight crew reported some malfunctions, but did not include the activation of stick shaker and STAB TRIM to CUT OUT. The AOA DISAGREE alert was not available on the aircraft therefore, the flight crew did not report it. The reported problem would only be able to rectify by performing tasks of AOA Disagree.

The following morning on 29 October 2019, the aircraft was operated from Jakarta with intended destination of Depati Amir Airport, Pangkal Pinang. According to the DFDR and the CVR, the flight had same problems as previous flight from Denpasar to Jakarta.

The flight crew started the IAS DISAGREE Non-Normal Checklist (NNC), but did not identify the runaway stabilizer. The multiple alerts, repetitive MCAS activations, and distractions related to numerous ATC communications contributed to the flight crew difficulties to control the aircraft.

The MCAS was a new feature introduced on the Boeing 737-8 (MAX) to enhance pitch characteristics with flaps up during manual flight in elevated angles of attack. The investigation considered that the design and certification of this feature was inadequate. The aircraft flight manual and flight crew training did not include information about MCAS.

On 10 March 2019, similar accident occurred in Ethiopia involved a Boeing 737-8 (MAX) experiencing erroneous of AOA.

As the result of the investigation safety actions have been taken by related parties. KNKT issued safety recommendations to address safety issues identified in this investigation to Lion Air, Batam Aero Technic, Airnav Indonesia, Boeing Company, Xtra Aerospace, Indonesia DGCA, and Federal Aviation Administration (FAA).

CHAPTER 1

FACTUAL INFORMATION

History of the Flight

A Boeing 737-8 (MAX) aircraft registered PK-LQP was being operated by PT. Lion Mentari Airlines (Lion Air) as a scheduled passenger flight from Soekarno-Hatta International Airport (WIII), Jakarta[1] with intended destination of Depati Amir Airport (WIPK), Pangkal Pinang[2]. The scheduled time of departure from Jakarta was 0545 LT on 29 October 2018 (2245 UTC[3] on 28 October 2018) as LNI610.
1
The number of persons onboard the aircraft was 189 consisted of two pilots, six flight attendants, and 181 passengers including one engineer. The weight and balance sheet showed the total person onboard was 188.

Prior to the departure, the Cockpit Voice Recorder (CVR) recorded the flight crew preflight briefing which mentioned Deferred Maintenance Item (DMI) of Automatic Directional Finder (ADF) that was unserviceable, the taxi route, runway in use, the intended cruising altitude of 27,000 feet and the weather being good. The CVR did not record the flight crew discussion related to the previous aircraft problem recorded in the Aircraft Flight and Maintenance Log (AFML).

The Captain acted as Pilot Flying (PF) and the First Officer (FO) acted as Pilot Monitoring (PM).

At 2315 UTC, the flight crew performed Before Taxi checklist. The Digital Flight Data Recorder (DFDR) recorded pitch trim was 6.6 units. Afterwards, the Jakarta Ground controller issued a taxi clearance to LNI610 flight crew and instructed to contact Jakarta Tower controller.

At 2318 UTC, the Jakarta Tower controller instructed the LNI610 flight crew to line up on runway 25 Left (25L). The flight crew then performed Before Takeoff checklist.

LNI610 flight crew and which was read back by the FO. At 23:20:01 UTC, the DFDR recorded Takeoff/Go-around (TO/GA) button was pressed and the engines spooled up to takeoff thrust.

At 23:20:16 UTC, the FO called 80 knots and the DFDR recorded the airspeed indicator on Captain's Primary Flight Display (PFD)[4] indicated 79 knots while on the First Officer's (FO) PFD indicated 81 knots. The DFDR also recorded difference angle between left and right Angle of Attack (AOA)[5] sensor, which was about 21° which continued until the end of recording. The DFDR indicated that the Flight/Director (F/D) on the Captain Primary Flight Display (PFD) showed 1° down, while on the first officer PFD showed 13° up.

At 23:20:32 UTC, the aircraft Enhance Ground Proximity Warning System (EGPWS) sounded "V1". The DFDR recorded the airspeed indicator on the Captain's PFD indicated 140 knots while on the FO's PFD indicated 143 knots. The low speed barber pole appeared on Captain's PFD airspeed indicator with the overspeed barber pole bar on the Vr (rotation speed) mark.

At 23:20:33 UTC, the FO called "rotate" and 2 seconds later as the nose gear lifted off the runway, the DFDR recorded left control column stick shaker[6] activation which continued for most of the flight.

At 23:20:37 UTC, the takeoff configuration warning sound was recorded momentarily on the CVR then the FO stating "Takeoff Config". Four seconds later, the Captain queried about the aircraft problem, at this time the DFDR recorded pitch was 7° up, the rate of climb was 1,000 feet/minute (fpm).

At 23:20:40 UTC, the aircraft became airborne. The clearance for departure was to follow the Standard Instrument Departure (SID) of ABASA 1C[7].

At 23:20:44 UTC, the FO called "Auto Brake Disarm" and advised the Captain of "Indicated Airspeed Disagree". The DFDR recorded the left indicated airspeed (IAS) was 164 knots and the right IAS was 173 knots. The IAS DISAGREE message appeared until the end of the recording. The FO then questioned what was the aircraft problem and asked whether the Captain intended to return to the airport. The Captain did not respond to the FO question and did not provide acknowledgement. The FO repeated the call "auto brake disarmed" which was acknowledged by the Captain.

At 23:20:51 UTC, the landing gear lever was moved to UP.

At 23:21:03 UTC, the Jakarta Tower controller instructed the LNI610 flight crew to contact Terminal East (TE) controller.

At 23:21:12 UTC, the FO advised the Captain "Altitude Disagree" and the Captain acknowledged. The altimeter on Captain's PFD indicated 340 feet and the FO's PFD indicated 570 feet.

At 23:21:22 UTC, the FO made initial contact with the TE controller who responded that the aircraft was identified on the controller Aircraft Situational Display/ASD (radar display). Thereafter, the TE controller instructed the LNI610 to climb to flight level 270.

At 23:21:28 UTC, the FO asked the TE controller to confirm the aircraft altitude as shown on the TE controller radar display. The TE controller responded that the aircraft altitude was 900 feet and was acknowledged by the FO. The DFDR recorded the altimeter on Captain's PFD indicated 790 feet and the FO's PFD indicated 1,040 feet.

At 23:21:37 UTC, the Captain instructed the FO to perform memory items for airspeed unreliable. The FO did not respond to this request.

At 23:21:44 UTC, the FO asked the Captain what the intended altitude he should request to the TE controller and suggested to the Captain to fly to downwind, which was rejected by the Captain. The Captain then commanded to the FO to request clearance to any holding point.

At 23:21:45 UTC, the DFDR recorded the aircraft started to turn to the left. The altimeter on Captain's PFD indicated 1,310 feet and on the FO's PFD indicated 1,540 feet. The heading bug was turned to the left.

At 23:21:52 UTC, the FO requested clearance from the TE controller "to some holding point for our condition now". The TE controller asked the

LNI610, what was the problem of the aircraft and the FO responded "flight control problem". The TE controller did not acknowledge the flight crew request to go to a holding point and only remembered the problem reported by the flight crew,

At 23:22:04 UTC, the FO suggested whether the Captain wanted to reconfigure the flaps setting to flaps 1, which the Captain agreed. The DFDR recorded that the flaps travelled from 5 to 1. About 10 seconds later, the Captain directed the FO to take over the control; the FO responded stating "standby".

At 23:22:15 UTC, the TE controller noticed that the LNI610 aircraft altitude on the radar display was decreasing from 1,700 to 1,600 feet and then the TE controller asked the intended altitude to the LNI610 flight crew. The DFDR recorded the altimeter on Captain's PFD indicated 1,600 feet and on the FO's PFD indicated 1,950 feet.

At 23:22:24 UTC, the FO suggested to the Captain to continue the flap reconfiguration and the Captain agreed. The flaps started to travel to UP position. The DFDR recorded the indicated airspeed on the Captain's PFD was 238 knots and the FO's PFD indicated 251 knots. Four seconds later, the FO asked the Captain if 6,000 feet would be the altitude they wanted, in response to the TE controller query. The Captain response was 5,000 feet.

At 23:22:30 UTC, the FO advised the TE controller that the intended altitude was 5,000 feet and the TE controller then instructed LNI610 to climb to an altitude of 5,000 feet and to turn left heading 050°. The instructions were acknowledged by the FO.

At 23:22:32 UTC, the aircraft EGPWS sounded: "BANK ANGLE, BANK ANGLE". The DFDR recorded the aircraft roll momentarily reached 35°.

At 23:22:33 UTC, the flaps reached the fully retracted position and the automatic AND trim was active for about 10 seconds, during which the horizontal stabilizer pitch trim decreased from 6.1 to 3.8 units.

At 23:22:41 UTC, the Captain instructed the FO to select flaps 1 and the DFDR recorded the flaps started to move. Three seconds later, the DFDR recorded the main electric trim moved the stabilizer in the aircraft nose up (ANU) direction for 5 seconds and the pitch trim gradually increased to 4.7 units.

At 23:22:44 UTC, the FO called 5,000. The Mode Control Panel (MCP) selected altitude began moving from 11,000 and reached 5,000 about 6 seconds later. At 23:22:45 UTC, the aircraft descended at a rate up to 3,570 fpm and lost about 600 feet of altitude. The DFDR recorded the pitch trim was at 4.4 units.

At 23:22:48 UTC, the flaps reached position 1 and the left control column stick shaker stopped briefly. The left AOA recorded 18° (nose up) and the right AOA recorded -3° (nose down). The rate of descent increased up to 3,200 fpm. On the Captain's PFD, the low speed barber pole appeared with the top of the pole was about 285 knots.

At 23:22:54 UTC, the automatic AND trim activated for 8 seconds at a low speed.

At 23:22:57 UTC, the FO asked the TE controller of the speed as indicated on the radar display.

At 23:23:00 UTC, the aircraft EGPWS sounded "AIR SPEED LOW – AIR SPEED LOW". The TE controller responded that the ground speed of the aircraft, shown on the radar display, was 322 knots. The DFDR recorded the indicated airspeed on the Captain's PFD indicated as 306 knots and on the FO's PFD indicated 318 knots.

At 23:23:00 UTC, the flight crew selected flaps 5 and the flaps began to travel from position 1 to 5. The Captain commanded ANU trim for 5 seconds and the pitch trim was recorded at 4.8 units. During this time, the AND automatic trim ended.

At 23:23:04 UTC, the left control column stick shaker activated and continued until the end of the recording, the rate of climb was about 1,500 feet/minute, and the pitch attitude was 3° nose up.

At 23:23:07 UTC, the DFDR recorded the flaps position was at 5.

At 23:23:08, the DFDR recorded on the Captain's PFD low speed barber pole and overspeed barber pole merged. On the FO's PFD, the overspeed barber pole appeared with the bottom of the pole about 340 knots and the low speed barber pole did not appear.

At 23:23:09 UTC, the Captain commanded "memory item, memory item".

At 23:23:15 UTC, the automatic AND trim activated for 1 second and activated again at 23:23:18 UTC for another 1 second.

At 23:23:17 UTC, the FO advised the Captain "Feel differential already done, auto brake, engine start switches off, what's the memory item here". The Captain then responded "check".

At 23:23:18 UTC, the automatic AND trim activated for 2 seconds and activated again at 23:23:23 UTC for another 2 seconds.

At 23:23:23 UTC, the automatic AND trim activated for 1 second and activated again at 23:23:26 UTC for 1 second, at this time the aircraft was on heading 100°.

At 23:23:26 UTC, the automatic AND trim activated for 1 second and activated again at 23:23:32 UTC for 2 seconds, at this time the aircraft stopped roll turn on heading 100°.

At 23:23:34 UTC, the FO asked "Flight control?" and the Captain responded "yeah".

At 23:23:39 UTC, the CVR recorded sound similar to paper pages being turned and the Captain commanded ANU trim for 1 second.

At 23:23:48 UTC, the FO called "flight control low pressure" and 4 seconds later the CVR recorded the sound of an altitude alert tone. At this time, the altimeter on the Captain's PFD indicated 4,110 feet and the FO's PFD indicated 4,360 feet. The automatic AND trim activated for 1 second and the aircraft began to turn to the left.

At 23:24:03 UTC, the DFDR recorded the Captain commanded ANU trim for 1 second.

At 23:24:05 UTC, the FO called "Feel Differential Pressure". Afterwards, the Captain commanded to perform the checklist for air speed unreliable, which was acknowledged by the FO.

At 23:24:27 UTC, the DFDR recorded the altimeter on the Captain's PFD indicated 4,900 feet and the FO's PFD indicated 5,200 feet. The aircraft climbed with a rate of about 1,600 feet/minute.

At 23:24:31 UTC, the FO advised the Captain that he was unable to locate the Airspeed Unreliable checklist.

At 23:24:43 UTC, the CVR recorded altitude alert tone and the DFDR recorded the altimeter on the Captain's PFD indicated 5,310 feet and the

FO's PFD indicated 5,570 feet with a rate of climb of about 460 feet/minute.

At 23:24:46 UTC, the CVR recorded a sound similar to paper pages being turned.

At 23:24:51 UTC, the TE controller added "FLIGHT CONT TROB" text for LNI610's target label on the controller radar system as a reminder that the flight was experiencing a flight control problem.

At 23:24:52 UTC, the DFDR recorded the flaps started retracting from 5 to 1. The CVR did not record any discussion related to flap position.

At 23:24:57 UTC, the TE controller instructed LNI610 flight crew to turn left heading 350° and maintain an altitude of 5,000 feet. The instruction was read back by the FO.

At 23:24:59 UTC, the DFDR recorded the flaps were at position 1. Four seconds later, the Captain commanded ANU trim for 4 seconds followed 1 second later by another ANU command for 1 second.

At 23:25:11 UTC, the FO repeated the TE controller instruction to the Captain to fly heading 350°, then informed him that there was no airspeed unreliable checklist.

At 23:25:13 UTC, the DFDR recorded that the flaps started traveling from position 1 to 0 (UP) and reached position 0 at 23:25:27 UTC. The CVR did not record any discussion related to flap position.

At 23:25:17 UTC, the FO stated "10.1" and began reading the Airspeed Unreliable checklist. Note: The Airspeed Unreliable checklist is on page 10.1 of the QRH.

At 23:25:27 UTC, the automatic AND trim activated by the Maneuver Characteristic Augmentation System (MCAS) for 2 seconds and was interrupted by the Captain who commanded ANU trim for 6 seconds. The pitch trim recorded 6.19 units.

At 23:25:40 UTC, MCAS activated for 6 seconds. The pitch trim recorded 4.67 units. This MCAS activation was interrupted when the Captain commanded ANU trim at 23:25:46 UTC for 7 seconds and again at 23:25:54 UTC for 1 second. The pitch trim recorded 6.27 units.

At 23:26:00 UTC, MCAS activated for 7 seconds and was interrupted at 23:26:06 UTC when the Captain commanded ANU trim for 6 seconds. The pitch trim recorded 5.59 units.

At 23:26:17 UTC, MCAS activated for 4 seconds until it was interrupted at 23:26:20 UTC when the Captain commanded ANU trim for 4 seconds. The pitch trim recorded 5.6 units

At 23:26:29 UTC, MCAS activated for 3 seconds until it was interrupted at 23:26:32 UTC when the Captain commanded ANU trim for 3 seconds. The pitch trim recorded 5.0 units.

At 23:26:32 UTC, the DFDR recorded the aircraft heading was 015° while the last instruction by the TE controller was fly heading 350°. While the FO was reading the Airspeed Unreliable checklist, the TE controller instructed to turn right heading 050° and maintain 5,000 feet. The instructions were read back by the FO.

At 23:26:45 UTC, MCAS activated for 3 seconds until it was interrupted at 23:26:48 UTC when the Captain commanded ANU trim for 6 seconds. The pitch trim was recorded as 5.83 units.

At 23:26:59 UTC, MCAS activated for 5 seconds until it was interrupted at 23:27:03, when the Captain commanded ANU trim for 6 seconds. Also, at 23:26:59, that the TE controller instructed LNI610 flight crew to turn right heading 070° to avoid traffic. The DFDR recorded the aircraft heading was 023° while the last instruction by the TE controller was fly heading 050°. The FO was still reading the Airspeed Unreliable checklist step noting that the flight path vector and pitch limit indicator may be unreliable and did not respond to the TE controller's instruction, thereafter, the controller called LNI610 twice and the FO responded at 23:27:13 UTC.

At 23:27:03 UTC, the flight crew commanded ANU trim for 7 seconds.

At 23:27:15 UTC, MCAS activated for about 5 seconds until it was interrupted at 23:27:19, when the flight crew commanded ANU trim for 5 seconds. The FO was reading the Airspeed Unreliable checklist while the TE controller instructed LNI610 to turn right heading 090°, which was acknowledged by the FO. The DFDR recorded the aircraft heading was 038° while the last instruction by the TE controller was fly heading 070°. A few seconds later, the TE controller revised the instruction to stop the turn and fly heading 070°, which was acknowledged by the FO.

At 23:27:19 UTC, the flight crew commanded ANU trim for 5 seconds.

At 23:27:29 UTC, MCAS activated for about 5 seconds until it was interrupted at 23:27:33 when the Captain commanded ANU trim for 6 seconds. The pitch trim was recorded as 5.5 units

At 23:27:44 UTC, MCAS activated for 4 seconds and at 23:27:48 UTC, the Captain commanded ANU trim for 4 seconds. The Captain commanded ANU trim again at 23:27:53 for 3 seconds. The pitch trim recorded 5.7 units.

At 23:27:58 UTC, the FO continued reading the Airspeed Unreliable checklist and informed the Captain that he would check the Performance Inflight, which was acknowledged by the Captain.

At 23:28:01 UTC, MCAS activated for about 7 seconds until it was interrupted at 23:28:07 when the Captain commanded ANU trim for 7 seconds. The pitch trim recorded 5.30 units.

At 23:28:09 UTC, the FO instructed the Flight Attendant (FA) via interphone to enter the cockpit.

At 23:28:15 UTC, the TE controller provided traffic information to the LNI610 flight crew, which was acknowledged by the FO and 1 second later the flight crew commanded ANU trim for 1 second.

At 23:28:18 UTC, the FA entered the cockpit and the Captain commanded to call the engineer to the cockpit. The FO also repeated the Captain's instruction for the FA to call the engineer.

At 23:28:22 UTC MCAS activated for 1 second and the flight crew commanded ANU trim for 3 seconds. The pitch trim recorded 5.4 units.

At 23:28:30 UTC MCAS activated for about 4 seconds until it was interrupted at 23:28:33 when the flight crew commanded ANU trim for 6 seconds. The pitch trim recorded 5.6 units.

At 23:28:41 UTC, the CVR recorded a sound similar to the cockpit door opening and 14 seconds later, the Captain asked someone "look what happened".

At 23:28:43 UTC, the TE controller instructed LNI610 flight crew to turn left heading 050° and maintain 5,000 feet. The instruction was acknowledged by the FO. The DFDR recorded the aircraft heading was

045° while the previous instruction by the TE controller was to fly heading 070°.

At 23:28:44 UTC, MCAS activated for about 4 seconds until it was interrupted at 23:28:48 when the Captain commanded ANU trim for 6 seconds. The pitch trim was recorded as 5.7 units.

At 23:28:58 UTC, there was communication between FAs on the interphone which discussed that there was a technical issue in the cockpit.

At 23:28:59 UTC, MCAS activated for about 6 seconds until it was interrupted at 23:29:04 when the flight crew commanded ANU trim for 5 seconds. The pitch trim was recorded as 5 units.

At 23:29:14 UTC, MCAS activated for about 7 seconds, until it was interrupted at 23:29:20 when the Captain commanded ANU trim for 3 seconds. Also, at 23:29:14, the FO confirmed the aircraft condition of landing gear up and altitude of 5,000 feet. Four seconds later, the CVR recorded sound similar to the altitude alert tone, at this time the DFDR recorded the altimeter on the Captain's PFD indicated 4,770 feet and the FO's PFD indicated 5,220 feet.

At 23:29:25 UTC, the flight crew commanded ANU trim for 6 seconds. The pitch trim was recorded as 5.4 units.

At 23:29:37 UTC, the TE controller questioned the LNI610 flight crew whether the aircraft was descending as the TE controller noticed that the aircraft was descending on the radar screen. The FO advised the TE controller that they had a flight control problem and were flying the aircraft manually.

At 23:29:38 UTC, MCAS activated for about 4 seconds until it was interrupted at 23:29:41 when the Captain commanded ANU trim for 7 seconds. The pitch trim was recorded as 5.6 units.

At 23:29:45 UTC, the TE controller instructed the LNI610 flight crew to maintain heading 050° and contact the Arrival (ARR) controller. The instruction was acknowledged by the FO. The DFDR recorded the aircraft heading was 059° while the last instruction by the TE controller was heading 050°.

At 23:29:53 UTC, MCAS activated for about 6 seconds until it was interrupted at 23:29:58 when the Captain commanded ANU trim for 3 seconds. The pitch trim was recorded as 4.5 units.

At 23:30:02 UTC, the FO contacted the ARR controller and advised that they were experiencing a flight control problem. The ARR controller advised the LNI610 flight crew to prepare for landing on runway 25L and instructed them to fly heading 070°. The instruction was read back by the FO. The DFDR recorded the aircraft heading was 054° while the last instruction by the TE controller was heading 050°. At 23:30:06 UTC, MCAS activated for about 2 seconds until it was interrupted at 23:30:07 when the Captain commanded ANU trim for 6 seconds. The pitch trim was recorded as 5.1 units.

At 23:30:18 UTC, MCAS activated about for 6 seconds until it was interrupted at 23:30:23 when the Captain commanded ANU trim for 9 seconds. The pitch trim was recorded as 5.5 units.

At 23:30:38 UTC, MCAS activated for about 5 seconds until it was interrupted at 23:30:42 when the Captain commanded ANU trim for 3 seconds. The pitch trim was recorded as 4.8 units.

At 23:30:48 UTC, the Captain asked the FO to take over control of the aircraft.

At 23:30:49 UTC, the FO commanded ANU trim for 3 seconds. At 23:30:54, the FO replied "I have control".

At 23:30:57 UTC, the Captain requested to ARR controller to proceed to ESALA[8] due to weather which was approved by the ARR controller.

At 23:31:00 UTC, the automatic AND trim activated for 8 seconds, the pitch trim changed from 5.4 to 3.4 units.

At 23:31:07 UTC, the FO stated "wah, it's very".

At 23:31:08 UTC, the FO commanded ANU trim for 1 second and the pitch trim changed to 3.5 units. Meanwhile the Captain advised the ARR controller that the altitude of the aircraft could not be determined due to all aircraft instruments indicating different altitudes. The Captain used the call sign of LNI650 during the communication. The ARR controller acknowledged then stated "LNI610 no restriction".

At 23:31:15 UTC, MCAS activated for about 3 seconds until it was interrupted at 23:31:17 when the FO commanded ANU trim for 1 second, the pitch trim changed to 2.9 units and the FO's column sensor force recorded 65 lbs. of back pressure.

At 23:31:19 UTC, the FO commanded ANU trim for an additional 4 seconds and the pitch trim changed to 3.4 units.

At 23:31:22 UTC, the Captain requested the ARR controller to block altitude 3,000 feet above and below for traffic avoidance. The ARR controller asked the intended altitude.

At 23:31:27 UTC, MCAS activated for 8 seconds, the pitch trim changed to 1.3 units and the FO's control column sensor force recorded 82 lbs.

At 23:31:33 UTC, the FO informed to the Captain that the aircraft was flying down. At this time the DFDR recorded that the aircraft pitch angle was -2° and the rate of descent was about 1,920 fpm.

At 23:31:35 UTC, the Captain responded to the ARR controller with "five thou". The ARR controller approved the flight crew request.

At 23:31:36 UTC, the FO exclaimed the aircraft was flying down which then the Captain responded: "it's ok".

At 23:31:36 UTC, the FO commanded ANU trim for 2 seconds and the pitch trim changed to 1.3 units.

At 23:31:43 UTC, MCAS activated for 4 seconds, the pitch trim changed to 0.3 units and the FO's control column sensor recorded 93 lbs.

At 23:31:46 UTC, the FO commanded ANU trim for 2 seconds, the altitude indicated on the Captain's PFD was 3,200 feet, the FO's PFD indicated 3,600 feet, and the rate of descent was more than 10,000 feet/minute.
At 23:31:51 UTC, the EGPWS audible alert "TERRAIN - TERRAIN" followed by "SINK RATE" were heard on the CVR as well as an overspeed clacker.

At 23:31:53 UTC, MCAS activated until the DFDR stopped recording at 23:31:54 UTC and the CVR stopped recording 1 second later.

The ARR controller attempted to contact the LNI610 flight crew twice with no response. At 23:32:20 UTC, LNI610 aircraft target disappeared on the ASD and changed to flight plan track. The ARR controller and TE controller attempted to contact the LNI610 four more times with no response.

The ARR controller then checked the last known coordinates of LNI610 and instructed the controller assistant to report the occurrence to the operation manager.

The ARR controller requested several other aircraft to hold over the last known position of LNI610 and to conduct a visual search of the area.

About 0005 UTC (0705 LT), a tugboat crew found floating debris at coordinate 5°48'56.04"S; 107°7'23.04"E which is about 33 Nm from Jakarta on bearing 056°. The debris was later identified as part of LNI610 aircraft.

Injuries to Persons and damage to Aircraft

Persons

All 181 passengers and 8 crewmembers perished in the accident. The Captain was Indian, one of the passengers was Italian and others were Indonesian citizen. The aircraft was destroyed.

Personnel Information
Captain:

Captain Gender	:	Male
Age	:	31 years
Nationality	:	India
Date of joining company	:	25 April 2011
License	:	ATPL
Date of issue	:	28 July 2016
Aircraft type rating	:	Boeing 737
Instrument rating validity	:	31 May 2019
Medical certificate	:	First class
Last of medical	:	5 October 2018

Validity	:	5 April 2019
Medical limitation	:	Flight crew shall wear corrective lenses
Last line check	:	19 January 2018
Last proficiency check	:	7 October 2018

Flying experience

Total hours	:	6,028 hours 45 minutes
Total on type (Boeing 737)	:	5,176 hours
Last 90 days	:	148 hours 15 minutes
Last 30 days	:	80 hours 5 minutes
Last 7 days	:	13 hours 15 minutes
This flight	:	About 11 minutes

Captain Training Record Summary

The investigation examined the Captain's training record since 2011, which consisted of general, proficiency, line and recurrent checks. The Captain passed all checks. The detail of the Captain's training record is available in the appendices of this report. Several remarks during the simulator proficiency check were as follows:

- 12 May 2015, in the assessment item of "stall on final approach", the remark was lack of appropriate technique that resulted in a second stick shaker activation.

- 25 May 2017, the remark was the Crew Resource Management (CRM) needed to be improved.

- 23 May 2018, in the assessment item "teamwork exercise" the remark was to use standard signal for effective communication and good teamwork during abnormal or emergency situation.

First Officer (FO)

Gender	:	Male
Age	:	41 years
Nationality	:	Indonesia
Date of joining company	:	31 October 2011
License	:	CPL
Date of issue	:	15 May 1997
Aircraft type rating	:	Boeing 737
Instrument rating validity	:	31 August 2019
Medical certificate	:	First Class
Last of medical	:	28 September 2018

Validity	:	28 March 2019
Medical limitation	:	Flight crew shall possess glasses that correct for near vision
Last line check	:	4 July 2017
Last proficiency check	:	25 August 2018

FO Training Record Summary

The detail of the FO's training record is available in the appendices of this report. Several relevant remarks during the simulator proficiency check are as follows:

- 25 June 2013, the check result was unsatisfactory. In the assessment item non-precision approach, the FO did not tune the localizer which resulted in after crossing the Final Approach Fix (FAF) he failed to descend. The assessor considered that it was caused by lack of situational awareness or judgment.

- On the assessment item missed approach it was remarked that he had a problem understanding the sequence of auto flight.

- 11 July 2013, the check result indicated remarks on 15 assessment items. On the assessment item "Basic Flight Characteristic" and "Approach to Stall" was remarked that recovery stall required more power and more force to maintain path.
  Another remark was on assessment item "Workload Management" which was remarked that the FO tended to have press-on-it-is[9].

- 14 July 2013, the new-hire training records indicated: need more detail on procedures, after 4th attempt of engine failure after takeoff the result was satisfactory, during single engine operation need improvement of rudder usage during power up or down, need more discipline to follow F/D during single engine non precision and precision approach. In general, "tends fixation so awareness less." Corrective training performed by briefing.

- 17 July 2013 simulator type rating check performed with satisfactory result.

- 2 June 2014, it was records indicated that the FO had "major problem" to focus on short-final, too rush, and needed gentle handling on control column with small correction for pitch and attitude and must be patient with the result. The FO also needed to manage stress while aircraft attitude was changing such as pitch due to external aspect (wind etc.).

- 6 July 2014, on assessment item "Area Departure or Standard Instrument Departure (SID)", it was noted that the FO missed the initial altitude for the SID. On the assessment item "Flight Management System (FMS)" the remark was the FO missed identifying the Non-Normal Checklist (NNC).

- On assessment item "Non-Precision Approach", the remark was the FO did not follow the approach profile and wrong decent target and vertical speed.

- 28 February 2016, on assessment item "Non-Precision Approach" the remark was that the FO had difficulty in maintaining the aircraft straight on final course.

- 6 August 2016, during the proficiency check on assessment item "N-1 ILS Approach", the remark was difficulties to control aircraft during manual flight.

- 6 August 2016, during the recurrent check, the overall result as satisfactory standard.

- 23 April 2017, the remark was "application exercise for stall recovery is difficult due to wrong concept of the basic principal for stall recovery in high or low level."

- 23 August 2017, during the proficiency check on assessment item "N-1 ILS Approach", the remark was too slow on scanning when on final approach.

- 8 April 2018, the remark was to improve the situational awareness.

- 25 August 2018, on assessment item "Start Malfunction" and "Situational Awareness" the remark was the EGT start exceeded 765°C instead of maximum value of 725°C.

Flight Attendants

All flight attendants held valid Flight Attendant Certificates with rating for Boeing 737 and valid medical examination certificates.

Air Traffic Controller

	Terminal East	Arrival
Gender	Male	Male
Age	63 years	50 years
Nationality	Indonesia	Indonesia
Year of joining company	2013	2013

License	:		ATC	ATC
Date of issue	:		1 February 2015	1 February 2015
Type rating	:			
			• •	• •
			Approach Control Surveillance	Approach Control Surveillance
			• •	• •
			Approach Control Procedural	Approach Control Procedural
Date of issue	:		17 June 2018	30 June 2018
Validity			17 December 2018	30 December 2018
Medical certificate	:		Third Class	Third Class
Last of medical	:		11 January 2018	19 April 2018
Validity	:		11 January 2019	19 April 2019
Medical limitation	:		Holder shall possess glasses that correct for	Holder shall wear corrective lenses
			Terminal East near vision	**Arrival**
ICAO Language Proficiency	:		Level 4	Level 5
Date of issue			28 July 2018	30 November 2015
Validity			28 July 2021	30 November 2020
Working time[10]				
Last 7 days	:		22 hours	
				• • 43 hours
				(office works acted as Operation Manager)
				• • 1 hour 30 minutes
				(as controller)
Last 24 hours	:		1 hours	1 hour 30 minutes
Duty time[11]				
Last 7 days	:		12 hours	1 hour 30 minutes
Last 24 hours	:		1 hour	1 hour 30 minutes

Aircraft Information

General

Registration Mark	PK-LQP
Manufacturer	Boeing Company
States of Manufacturer	United States of America
Type/Model	737-8 (MAX)
Serial Number	43000
Year of Manufacture	2018
Certificate of Airworthiness	
Issued	15 August 2018
Validity	14 August 2019
Category	Transport
Limitations	None
Certificate of Registration	
Number	43000
Issued	15 August 2018
Validity	14 August 2021
Time Since New	895 hours 21 minutes
Cycles Since New	443 cycles
Last Major Check	None
Last Minor Check	None

Engines

Manufacturer:	CFM International
Type/Model:	LEAP n- 1B25

Serial Number 1 engine: 602506
- Time since new: 895 hours 21 minutes
- Cycles since new: 443 cycles

Serial Number 2 engine: 602534
- Time since new: 895 hours 21 minutes
- Cycles since new: 443 cycles

Recorded Aircraft Problems

The recorded aircraft problems based on Onboard Maintenance Function (OMF) and Aircraft Flight and Maintenance Log (AFML) are as follow.

9 October 2018

The problems related to the air data system were recorded intermittently in the Flight History page of the aircraft OMF. The OMF Flight History page recorded that at 1159 UTC the message "Angle of Attack Signal is Out of Range" was detected by the left Air Data Inertia Reference Unit (ADIRU).

The AFML has an entry that on 9 October 2018 at 1219 UTC, during parking in Jakarta the aircraft had the problem of STBY PWR OFF (stand by power off) light illuminated and followed by the Circuit Breakers (CBs) tripped out on the following components; DC battery, APU GCU (Auxiliary Power Unit Generator Control Unit), GCU 1 (Generator Control Unit of left engine) and GCU 2 (GCU right engine), and GEN DISC 1 (Generator Disconnect of the left generator) and GEN DISC 2 (Generator Disconnect of the right generator).

The engineer in Jakarta rectified the problem by closing all the CBs and conducted engine run at idle power to verify the problem. The AFML entry stated that the problem was resolved.

26 October 2018
The aircraft arrived in Manado at 2243 UTC after a flight from Tianjin (China) with flight number LNI2748. The AFML entry states the SPD (Speed) and ALT (Altitude) Flags appeared on Captain's PFD which means flags appear instead of normal airspeed and altitude indication. The MAINT light on the overhead panel illuminated. The engineer in Manado checked the OMF and found the maintenance message 27-31000 (SMYD FAULT, detected by Stall Management Yaw Damper (SMYD) - 1).

Regarding maintenance message 27-31000, using the Interactive Fault Isolation Manual (IFIM) led to two maintenance messages which were "Spoiler Load Alleviation" with the IFIM task 27-60-00-810-801 and "SMYD Fault" with IFIM task 27-32-00-810-805.

The engineer in Manado performed IFIM task 27-32-00-810-805 by conducting self-test of SMYD position 1. The AFML entry states that the self-test result passed and the engineer erased the maintenance message. The airspeed and altitude indicators appeared on captain's PFD and the engineer in Manado released the aircraft for flight.

26 October 2018
The aircraft flew from Menado on 26 October 2018 (UTC) to Denpasar (LNI775) arrived on 27 October 2018 (UTC), the SPD and ALT flags appeared on the Captain's PFD were reported occurred on this flight. The engineer in Denpasar checked the OMF and found the maintenance

message 27-31012 (AD DATA INVALID detected by Stall Management Yaw Damper (SMYD) -1). Afterward, the engineer in Denpasar performed the SMYD-1 self-test which result passed. Subsequently the engineer in Denpasar erased the message in the OMF which resulted in the MAINT light on the P5 panel was extinguished.

The aircraft and the flight crew then performed a round trip flight, flying Denpasar to Lombok (LNI828), and Lombok to Denpasar (LNI829). The SPD and/or ALT flags were not reported as having occurred on these flights.

The aircraft then flew from Denpasar to Manado (LNI776) and arrived in Manado at 0845 UTC. The AFML (page number B3042851) recorded the indication of the SPD and ATRIM and MACH TRIM lights illuminated. .

The aircraft was scheduled to stay over-night in Manado before scheduled for flight to Denpasar on the following morning. At the time of the problem rectification, the weather at Manado was raining.

The AFML record the engineer referenced IFIM task 27-32-00-810-816 (AD DATA INVALID) and reviewed the OMF which showed formed the self-test of the SMYD-1, which resulted in a failure indication.

The engineer in Manado found the correlated maintenance messages on the OMF which were 27-31012 (AD DATA INVALID), 34-21107 (ADIRU-L ADR DATA SIGNAL IS INVALID), 34-21123 (ANGLE OF ATTACK SIGNAL IS OUT OF RANGE) and 34-61263 (AIR DATA SIGNALS FROM ADIRU-L ARE INVALID).

The OMF maintenance message 27-31012 led the engineer to IFIM task 27-32-00-810-816 (AD DATA INVALID) which required the self-test of the SMYD-1 which had already been done.

The OMF maintenance message 34-21107 and 34-21123 led to conducting the Build in Test Equipment (BITE) test of the system through FMC CDU (Flight Management Computer - Computer Display Unit).

The engineer in Manado conducted the FMC CDU BITE test and found the maintenance message 34-21007 (AIR DATA INVALID) and 34-21023 (AOA SIGNAL FAIL).

Subsequently the engineer in Manado reset the CB of AC, DC and EXC (excitation power source) of the left ADIRU (Air Data Inertial Reference Unit). After resetting the CBs, the Manado engineer conducted self-test of the SMYD-1 and Digital Flight Control System (DFCS) with result that it passed.

The OMF maintenance message 34-61263 led to the IFIM task 34-61-00-810- 854 (AIR DATA SIGNALS FROM ADIRU-L ARE INVALID, detected by flight management computer (FMC)-1) which consists of 14 steps in the Fault Isolation Procedure including the wiring check of Air Data Module (ADM) and ADIRU. The wiring checks were not conducted by the engineer in Manado because the weather was raining and to avoid lightning hazard. The engineer in Manado performed the electrical connector inspection by removing and reconnecting the connectors of the left ADM and ADIRU. The inspection did not find abnormality to the left ADM and ADIRU.

The AFML recorded that the fault status of OMF maintenance message regarding the problems were not active.

On the following morning before the flight, the engineer in Manado met with the flight crew and discussed the rectification of the SPD and ALT flags problem. The flight crew of this departing flight was the same flight crew of the inbound flight on this aircraft on the day before. The flight crew mentioned that the problem appeared several times and requested more be done to rectify the problem. The engineer suggested that it might be better to conduct problem solving in Denpasar.

28 October 2018
The aircraft departed Manado on 27 October 2018 at 2340 UTC (28 October 2018 at 0740 LT) with the flight number LNI775 and arrived at Denpasar on 28 October 2018 at 0205 UTC. The AFML (page number B3042853) recorded the problem as follows:

1. SPD and ALT Flags shown on Captain's PFD,

2. SPEED TRIM and MACH TRIM lights illuminated,

3. Auto throttle arm disconnect on take-off roll.

The AFML entry referred to IFIM task 27-32-00-810-816 (AD DATA INVALID) and referred to the OMF which showed the "STALL WARNING SYS L" on the status message page. The engineer in Denpasar performed self-test of the SMYD-1 with result that it failed. The OMF showed maintenance messages of 27-31012 (AD DATA INVALID) and 27-31015 (ADIRU-L IR1 DATA INVALID – Inertial Data is Invalid).

The engineer in Denpasar checked the existing fault in the OMF and found maintenance message of 34-21107 (ADIRU-L ADR DATA SIGNAL IS INVALID) and 34-21123 (AOA SIGNAL OUT OF RANGE).

The engineer conducted BITE test via FMC CDU and found maintenance messages 34-21007 (ADR DATA INVALID) and 34-21023 (AOA SIGNAL FAIL).

Subsequently the engineer in Denpasar reset the CB of AC, DC and EXC (excitation power source) of the left ADIRU. After resetting the CBs, the engineer conducted self-test with result that it passed.

The engineer in Denpasar also conducted the BITE test of the Digital Flight Control System (DFCS) which result that it passed and in the OMF showed the maintenance message not active. Thereafter, the engineer erased the status message.

The AFML entry stated the engineer in Denpasar intended to replace the AOA sensor for trouble shooting due to repetitive problem.

Because a spare AOA sensor was not available in Denpasar, the engineer coordinated with Maintenance Control Centre (MCC) to order an AOA sensor from Batam Aero Technique (BAT) located in Batam. While waiting for the AOA sensor to arrive in Denpasar, the aircraft was grounded.

The AOA sensor (part number 0861FL1 serial number 14488) arrived in Denpasar about 1000 UTC (1800 LT). The engineer in Denpasar removed the AOA sensor (part number 0861FL1 serial number 21401) referring to AMM 34-21-05-000-801 and installed the AOA received from BAT, referring to AMM 34-21-05-400- 801. After installation of the AOA, the AMM requires performing an installation test by referring to task AMM 34-21-05-400-801. The AMM described two methods for performing the installation test; the recommended method by using test equipment AOA test fixture SPL-1917 and the alternative method using the SMYD BITE module. The test equipment (AOA test fixture SPL-1917) was not available in Denpasar therefore, the engineer in Denpasar used the alternative method. The alternative method is performed by deflecting the AOA vane to the fully up, center, and fully down positions while verifying the indication on the SMYD computer for each position. The engineer did not record the indication on the SMYD computer during the installation test.

The engineer in Denpasar conducted the heater test by dropping water onto the AOA vane with result that it passed. The engineer performed the BITE test on the FMC CDU which showed "No Current Faults".

The engineer in Denpasar provided the investigation several photos including of the Captain's PFD that was claimed to be taken after the AOA sensor replacement and of the SMYD during the installation test. However, the time shown on the Captain's PFD was the time before arrival of AOA sensor spare part and the investigation confirmed that the SMYD photos were not of the accident aircraft.

The aircraft was released to service at 1230 UTC (2030 LT).

Thereafter, the aircraft departed from Denpasar to Jakarta with flight number LNI043 and arrived in Jakarta at 1556 UTC (2256 LT). The flight crew reported on AFML page number B3042855 that the aircraft had problems of "IAS and ALT Disagree shown after take-off" and "FEEL DIFF PRESS light illuminated".

The AFML recorded that engineer in Jakarta conducted flushing on left pitot and static ADM refer to the IFIM task 34-20-00-810-801 (ALT DISAGREE Shows on entry stated the of the PFD (Captain's)- Fault Isolation) and conducted the operational test which resulted in satisfactory performance. The AFML page number B3042855 also recorded that the Jakarta engineer cleaned the electrical connector of the elevator feel computer and referred to the IFIM task 27-31-00-810-803 (FEEL DIFF PRESS Light is on-Fault Isolation) and conducted the test with result that it passed.

The engineer released the aircraft into service on 28 October 2018 at 1930 UTC (29 October 2018 at 0230 LT).

Angle of Attack (AOA) Sensors Historical Record

Removed Angle of Attack sensor (P/N 0861FL1; S/N 21401)
Lion Air removed AOA sensor Part Number (P/N) 0861FL1 Serial Number (S/N) 21401 from PK-LQP aircraft on 28 October 2018 to address a maintenance write-up stating that the speed (SPD) and altitude (ALT) flags appeared on the Captain's PFD. Following the accident, BATAM Aero Technik provided the removed AOA sensor to KNKT on 5 November 2018. The KNKT subsequently provided the sensor to the United States National Transportation Safety Board (NTSB) for further examination and detailed testing.

On 10 December 2018, representatives from the KNKT, NTSB, FAA, and Boeing, convened at a Collins Aerospace (previously known as Rosemount Aerospace) facility to perform examination and testing of the AOA sensor in accordance with the Collins Aerospace Component Maintenance Manual (CMM) 34-12-34, Revision 9. Examination of the AOA sensor revealed an intermittent open circuit in the resolver #2 coil wiring. At temperatures above approximately 60°C, the resolver functioned normally, but did not function below that temperature.

The detail report of the inspection result of AOA sensor P/N 0861FL1 S/N 21401 is attached in the appendices of this report.

Installed Angle of Attack (AOA) sensor (P/N 0861FL1; S/N 14488)

At the time of the accident, AOA sensor, P/N 0861FL1, S/N 14488 was installed on the left side of the fuselage of PK-LQP; this sensor had been installed on 28 October 2018 in Denpasar. No fragments of the AOA sensor were identified in the recovered wreckage.

This AOA sensor was previously installed on the right side of the fuselage of a Boeing 737-900ER aircraft, Malaysian registration 9M-LNF, which was operated by Malindo Air. The AOA sensor had been removed on 19 August 2017 due to maintenance write-ups indicating that SPD and ALT flags were shown on the FO PFD during a pre-flight check.

The investigation reviewed Flight Operation Quality Assurance (FOQA) data of 9M-LNF aircraft from 28 May 2017 to 26 August 2017. On the Boeing 737-900ER, airspeed, altitude, and AOA from the right side are not recorded on the DFDR. However, the DFDR does record Flight Director (F/D) commands on the FO PFD which are dependent on airspeed and altitude measurements. The review found 15 flights that experienced the right F/D parameter recorded as "no computed data" consistent with the appearance of the SPD and ALT flags on the FO's PFD.

On 22 August 2017, the AOA sensor S/N 14488 was sent to, and subsequently stored at a Malindo Store & Logistic Department in Kuala Lumpur, Malaysia. A Repair Order was issued by Batam Teknik (Batam Aero Technic - BAT) on 23 August 2017.

On 20 October 2017, Batam Teknik sent S/N 14488 from Kuala Lumpur to Xtra Aerospace in Miramar, Florida, USA for repair.

From 23 October 2017 to 3 November 2017 S/N 14448 was under repair at the Xtra Aerospace facility. Work Order number W8206 noted that the reason of removal was speed (SPD) and altitude (ALT) flags displayed and speed and altitude indication did not appear.

Preliminary inspection of the AOA sensor at Xtra Aerospace verified the part number and serial number. The unit was observed to be in fair but dirty condition and did not pass the operational test. The preliminary result stated that the eroded vane caused erroneous readings.

Based on repair records, the unit was disassembled to replace the eroded vane. After vane replacement and reassembly the unit was calibrated and tested to the requirements of the CMM Revision 8 (current at the time of the repair). The work order stated that the results for the required tests were satisfactory. Xtra Aerospace approved S/N 14488 for return to service on 3 November 2017.

On 27 November 2017, Xtra Aerospace sent the unit to the Malindo Air stores facility where it was received in Kuala Lumpur on 01 December 2017. Malindo Air sent S/N 14488 from Kuala Lumpur to BAT store in Batam on 20 December 2017. BAT received S/N 14488 on 22 December 2017.

On 28 October 2018, the AOA sensor sent from Batam at 0900 LT (0200 UTC) and received at Denpasar station about 1830 LT (1030 UTC) where it was installed on the left position of PK-LQP aircraft.

Aircraft System information
These aircraft system descriptions are referring to Boeing 737-8 (MAX) Aircraft Maintenance Manual (AMM), System Description Section (SDS) and Flight Crew Operations Manual (FCOM).

Air Data Inertial Reference System
The Boeing 737-8 (MAX) is equipped with an Air Data Inertial Reference System (ADIRS) that provides flight data to the flight deck display panels, flight management computers, flight controls, engine controls and all other systems requiring inertial and air data information. The ADIRS combines the Air Data System (ADS) function and the Inertial Reference System (IRS) function into a single device identified as an Air Data Inertial Reference Unit (ADIRU). The ADIRUs provide inertial position and track data to the flight management system as well as attitude, altitude and airspeed data to the flight deck displays. The ADIRUs process information measured by internal gyros and accelerometers and information from the air data sensors.

Air Data System (ADS)

The function of the Air Data Module (ADM) is to sense the aircraft total and static pressures external to the aircraft and convert them into digital electrical signals. These electrical pressure signal, in conjunction with the Total Air Temperature RU to calculate basic air data information (parameters) for transmission to various systems on the aircraft. Some of the parameters that the ADIRU transmits include: altitude, computed airspeed, and true airspeed. Another function of the ADIRU is to provide AOA information (corrected angle of attack) directly to the Flight Control Computers (FCC) as in input to the MCAS function.

Both the altitude and airspeed use static pressure which includes calculations for a correction factor of the Static Source Error Correction (SSEC). This is a compensation for pressure errors caused by the airframe aerodynamic effects on the static port. The static ports have been located to minimize errors. Compensation for the remaining errors is provided by a correction algorithm comprised of three factors: basic correction, thrust effect compensation and ground effects compensation

Pitot and Static System

The pitot static system is comprised of three separate pitot probes and six flush static ports; two of these pitot probes and four of the static ports interface with the Air Data Modules (ADM), which convert pneumatic pressure to electrical signals and send these data to the ADIRUs. The ADM component is installed inside the pressurized cabin. The cabin altitude is normally rate–controlled by the cabin pressure controller up to a cabin altitude of 8,000 feet at the aircraft maximum certified ceiling of 41,000 feet.

The remaining auxiliary pitot probe and alternate static ports provide pitot and static pressure to the standby instruments. The auxiliary pitot probe is located on the first officer's side of the aircraft.

The ADM connected to the Captain's pitot probe sends information to the left ADIRU, while the ADM connected to the First Officer's pitot probe sends information to the right ADIRU. The remaining ADMs are located at the balance centers of the Captain's and First Officer's static ports. The ADM connected to the Captain's static ports sends information to the left ADIRU, while the ADM connected to the First Officer's static ports sends information to the right ADIRU.

The data from the ADIRU is processed by the Display Processing Computer (DPC) in the MAX Display System (MDS). The Boeing 737-8 (MAX) has two DPCs. The DPC receives ARINC 429 digital data and analog discrete from various aircraft systems. The aircraft systems send sets of data to DPC 1 and DPC 2. The DPCs processes these data to be displayed on the Display Units (DU) located within the flight deck.

In the event of certain system failures, the ADIRU output data provided to other systems, including the DPC, may become invalid (No Computed Data (NCD) or Failure Warning (FW). In response, DPC and the Primary Flight Display (PFD) will show a flag on the particular parameter (ALT, SPD, ATT, etc.) with amber color and the particular parameter will not be shown in the PFD.

The ALT and/or SPD flags will appear on the PFD if the altitude and/or computed airspeed data from the ADIRU is invalid. The respective altitude and/or computed airspeed data will not be shown on the PFD. The parameter flags that appear on the PFD are shown in the figure below.

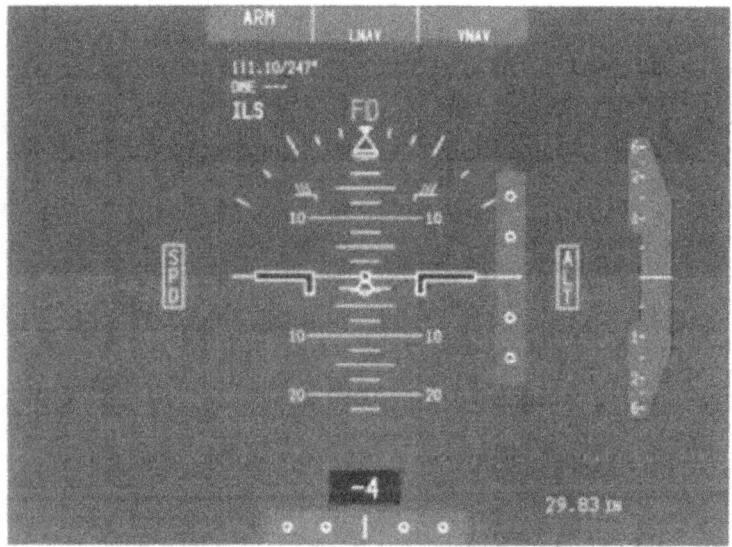

Figure 1. Instrument SPD and ALT flags appear on PFD

SPD Flag (amber) means the computed airspeed indication is inoperative.

In the DFDR, there is no discrete parameter indicating that the SPD flag is being displayed. However, if either the DFDR parameter of computed airspeed (CAS) left or right shows the characteristic "saw tooth" pattern, that indicates an invalid data status which can be concluded that the SPD flag is being displayed on Captain or First Officer PFD.

ALT Flag (amber) means the altitude display has failed. In the DFDR, there is no discrete parameter indicating that the ALT flag is being displayed. However, the DFDR records four parameters of barometric altitude on each altimeter. If these altitudes are marked by the ADIRU as invalid, the DFDR records a "saw tooth" error pattern from which it can be concluded that the ALT flag is being displayed on captain's or FO's PFD.

Both DPCs compare each other's data and in the case that the data is not similar at certain values for a certain period of time, the corresponding disagree message will be displayed on both PFDs.
1. IAS disagree (Indicated Airspeed disagree) message appears if the airspeed indications on both PFDs different by more than 5 knots for more than 5 seconds.
2. ALT disagree (altitude disagree) message appears if the altitude indication on both PFDs different by more than 200 feet for more than 5 seconds.

Airspeed Low Alert

The AIRSPEED LOW annunciation alerts flight crew for low air speed. The alert is an aircraft operational alert that is calculated by the Enhance Ground Proximity Warning System (EGPWS) and the MAX Display System (MDS) which occurs when the computed airspeed (from the ADIRU) falls below a threshold airspeed between the minimum maneuver speed and stick shaker speed.

The aural coincides with the low airspeed alert on the airspeed indication. The minimum maneuver speed is indicated by amber bar on the PFD with the first flap retraction after takeoff or when valid Vref is entered.

Top of amber bar indicates minimum maneuver speed. This airspeed provides:

- The 1.3 g maneuver capability to stick shaker below approximately 20,000 feet.
- The 1.3 g maneuver capability to low speed buffet (or an alternative approved maneuver capability set in the FMC maintenance pages) above approximately 20,000 ft.

The minimum speed indicated in red and black (barber pole). Top of bar indicates the speed at which stick shaker occurs.

Maximum Operating Speed (MMO or VMO):

Boeing 737-8 (MAX) presented the Maximum Operating Speed (Maximum Mach Operating Speed (Mmo) or Maximum Operating Speed (Vmo)) in the red and black (barber pole) warning bands and Maximum Maneuver Speed presented in amber bar on top of speed tape indication on the PFD. The bottom of the barber pole bar indicates the maximum speed as limited by the lowest of the following:

- Vmo/Mmo
- Landing gear placard speed
- Flap placard speed.

When either an over-speed condition or a system test occurs, a clacker aural warning will active. The warning clackers can be silenced only by reducing airspeed below Vmo/Mmo. The over-speed warning system can only be tested on the ground.

Angle of Attack (AOA) Sensors

The Boeing 737-8 (MAX) has two independent angle-of-attack (AOA) sensors, one on each side of the forward fuselage. The AOA sensors consist of an external vane which rotates to align with the local airflow connected to two internal resolvers which independently measure the rotation angle.

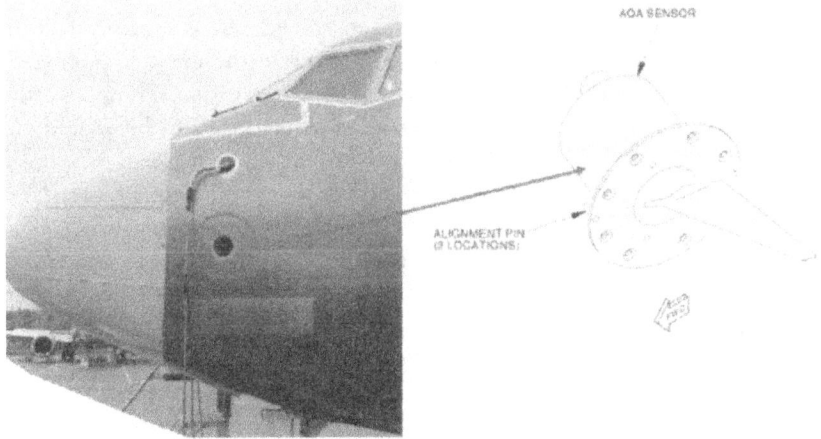

Figure 2: Angle of Attack (AOA) Sensor

The AOA sensor used on the Boeing 737-8 (MAX)-8 is made by Collins Aerospace (previously Rosemount Aerospace).

For each AOA sensor (left and right), one resolver is connected to the respective Stall Management Yaw Damper (SMYD) computer and the second resolver is connected the respective ADIRU. Both the SMYD and ADIRU monitor the resolver circuits within the AOA sensor. If a fault is detected, the AOA resolver information is not used and the fault is annunciated.

There is no scheduled maintenance for AOA sensors. Any required maintenance is a consequence of annunciated faults or observed malfunctions. This practice is known as "on-condition" maintenance.

AOA Display Option

Boeing provides the option for the operator to install the AOA indicator on the PFD for Boeing 737-8 (MAX).

The "AOA DISAGREE" message appears on the Captain and First Officer PFD when the values of the left and right AOA transmitted by the ADIRUs differ by 10° or more for 10 continuous seconds. The annunciation is only displayed in the air because AOA values are unreliable when the aircraft is stationary on the ground.

The AOA DISAGREE message was first implemented on the Boeing 737 NG fleet in 2006 in response to customer requests. Since 2006, the AOA DISAGREE alert has been installed on all newly manufactured Boeing 737 NG aircraft, and is available as a retrofit for older aircraft.

The AOA DISAGREE alert has not been considered as a safety feature by Boeing, and is not necessary to safely operate the aircraft. Airspeed, attitude, altitude, vertical speed, heading and engine thrust settings are the primary parameters the flight crews use to safely operate the aircraft in normal flight. Stick shaker and the pitch limit indicator are the primary features used for the operation of the aircraft at elevated angles of attack. The AOA DISAGREE alert provides supplemental information only. The AOA DISAGREE non-normal procedure alerts pilots to the possibility of airspeed and altitude errors, and of the IAS DISAGREE and ALT DISAGREE alerts occurring; but the non-normal procedure does not include any flight crew action in response to the AOA DISAGREE alert.

The requirements for the AOA DISAGREE alert were carried over from the Boeing 737 NG to the Boeing 737-8 (MAX). In 2017, however, within several months after beginning Boeing 737- 8 (MAX) deliveries, Boeing identified that the Boeing 737-8 (MAX) display system software did not correctly implement the AOA DISAGREE alert requirements. As with the Boeing 737 NG, the Boeing display system requirements for the Boeing 737-8 (MAX) called for the activation of the AOA DISAGREE alert as a standard feature on all aircraft. The software delivered to Boeing, however, linked the AOA DISAGREE alert to the AOA position indicator, which is an optional feature on the Boeing 737 (MAX) series. Accordingly, the software activated the AOA DISAGREE alert only if an airline opted for the AOA indicator. At the time of the accident, Boeing advised that the AOA indicator has been selected by approximately 20% of airlines.

When the discrepancy between the AOA display requirements and the software was identified, Boeing determined that the absence of the AOA DISAGREE alert did not adversely impact aircraft safety or operation. Accordingly, Boeing concluded that the existing functionality was acceptable until the originally intended functionality could be implemented in a display system software upgrade, scheduled for the third quarter of 2020.

Lion Air did not select the optional AOA indicator feature on the PFD of their 737-8 (MAX) aircraft. As a result, the AOA DISAGREE did not appear on PK-LQP aircraft, even though the necessary conditions were met.

Following the Lion Air accident, Boeing convened a Safety Review Board (SRB) to reconsider whether the absence of the AOA DISAGREE alert from certain Boeing 737-8 (MAX) flight displays presented a safety issue. That SRB confirmed Boeing's prior conclusion that it did not. Boeing also elected to accelerate the software change. Boeing advised that new software implementing the AOA DISAGREE alert will be available before the Boeing 737-8 (MAX) aircraft return to service.

All customers with previously delivered Boeing 737-8 (MAX) aircraft will have the ability to activate the AOA DISAGREE alert per a service bulletin to airlines.

Horizontal Stabilizer

Pitch control for the Boeing 737-8 (MAX), is provided by two elevators and a movable horizontal stabilizer.

The Horizontal Stabilizer consists of a left, right, and center section. It pivots at a hinge point located at the aft end of the center section. The stabilizer moves to a maximum of 4.2 degrees leading-edge up and 12.9 degrees leading-edge down.

The total range of the Horizontal Stabilizer movement is 17.1 degrees (or units) which is depicted on the scale on the stabilizer trim indicator located on the center pedestal in the cockpit. When the stabilizer trim indicator is at the 0 position, the Horizontal Stabilizer is at the full leading-edge up position. This will mean the aircraft is trimmed in a full nose-down position. When the horizontal stabilizer is at a position of 0 degrees relative to the aircraft centerline, the stabilizer trim will indicate 4 units (Figure 3).

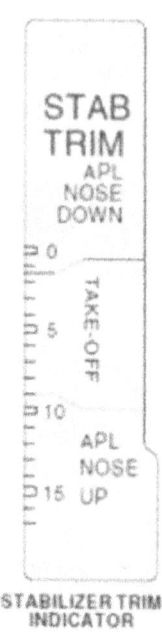

Figure 3: The stabilizer trim indicator

The Horizontal Stabilizer pivots through about 17 degrees and the position is shown on the flight deck stabilizer trim indicator. Lower values indicate more nose-down trim and high values indicate more nose-up trim. Before takeoff, the flight crew set the position of the horizontal stabilizer based on the location of the aircraft's center of gravity.

The horizontal stabilizer is positioned by a single electric trim motor controlled through either the main electric trim switches on the control wheels or automatic trim. The stabilizer may also be positioned by manually rotating the stabilizer trim wheels which are located on either side of the aisle stand.

Main electric trim switches on each control wheel actuate the electric trim motor through the main electric stabilizer trim circuit when the aircraft is flown manually. Automatic trim is accomplished through the autopilot stabilizer trim circuit. The main electric and autopilot stabilizer trim have two rates: high trim rate with flaps extended and low trim rate with flaps retracted. Actuating either pair of main electric trim switches will automatically disengage the autopilot (if engaged), and will override any automatic trim commands. The stabilizer trim wheels rotate whenever the stabilizer is in motion.

Main electric trim switches on each control wheel actuate the electric trim motor through the main electric stabilizer trim circuit when the aircraft is flown manually. Automatic trim is accomplished through the autopilot stabilizer trim circuit. The main electric and autopilot stabilizer trim have two rates: high trim rate with flaps extended and low trim rate with flaps retracted. Actuating either pair of main electric trim switches will automatically disengage the autopilot (if engaged), and will override any automatic trim commands. The stabilizer trim wheels rotate whenever the stabilizer is in motion.

The STAB TRIM PRI (stabilizer trim primary) cutout switch and the STAB TRIM B/U (stabilizer trim back up) cutout switch are located next to each other on the aisle stand just aft of the thrust levers. If either switch is positioned to CUTOUT, power is removed from the stabilizer trim motor and neither main electric trim nor automatic trim can move the stabilizer.

The stabilizer control system prevents the stabilizer from moving in opposition to the control column. For example, if the column is being pulled aft (to pitch up), the stabilizer is prevented from trimming in the nose-down direction. The addition of MCAS to the Boeing 737-8 (MAX) required a modification of this function.

Feel Differential Pressure

The elevator feel computer provides simulated aerodynamic forces on the control column using total pressure from two dedicated pitot probes mounted on the vertical stabilizer and stabilizer position. Feel force is transmitted to the control columns by the elevator feel and centering unit - thus column forces are adjusted relative to the airspeed. Elevator Feel Shift (EFS) modifies the column forces at high angles-of-attack.

The FEEL DIFF PRESS light (Feel Differential Pressure) on the overhead panel will illuminate if EFS operates continuously for more than 30 seconds. The FEEL DIFF PRESS light is not recorded on the DFDR, but will result in activation of Master Caution which is recorded on the DFDR.

Speed Trim System (STS)

The Speed Trim System (STS) provides speed stability augmentation and pitch stability augmentation. Speed stability augmentation is provided by the Speed Trim Function. Pitch stability augmentation is provided by the MCAS function.

The Speed Trim Function is designed to improve flight characteristics during operations at low gross weight, aft center of gravity and high thrust when the autopilot is not engaged. The Speed Trim Function operates most frequently during take-off, climb and go-around. The flight directors may be on or off.

The Speed Trim Function helps keeps the speed set by the pilots with commands to the horizontal stabilizer. As the aircraft speed slows, the stabilizer is moved to a more nose down position to increase the speed. As the speed increases, the stabilizer is moved to a more nose up position to decrease the speed.

MCAS Functional – Detailed Description

The MCAS is a function within the Speed Trim System and, when activated, moves the stabilizer during non-normal flaps up, high angle of attack maneuvers to provide a desirable increase in stick force gradient and a reduced pitch up tendency. Similar to the Speed Trim Function, the MCAS function is also a flight control law contained within each of the two FCCs. MCAS is only active in the master FCC for that flight. At aircraft power-up, the master FCC defaults to the left side FCC; and will then alternate between the left and right FCC by flight. The master FCC is not affected by the position of the Flight Director switches. The FCCs receive inputs from several systems including the air data inertial reference system (ADIRS). Specific to the MCAS, the control law commands the stabilizer trim as a function of the following: Air/Ground, Flap position, Angle of attack, Pitch rate, True Airspeed and Mach.

The AOA and Mach inputs are provided to each FCC by the associated Air Data Inertial Reference Unit (ADIRU). Each ADIRU receives AOA information from one of the two resolvers contained within the associated AOA sensor (i.e. the Left ADIRU uses left AOA vane and the Right ADIRU uses the right AOA vane). Information from the other resolver contained within the AOA sensor, along with data from other sources, is provided to the Stall Management Yaw Damper computer (SMYD), which is used, along with data from other sources, for the purpose of calculating and sending commands to the Stall Warning System (SWS).

As originally delivered, the MCAS became active during manual, flaps-up flight (autopilot not engaged) when the AOA value received by the master

FCC exceeded a threshold based on Mach number. When activated, the MCAS provided a high rate automatic trim command to move the stabilizer AND. The magnitude of the AND command was based on the AOA and the Mach. After the non-normal maneuver that resulted in the high AOA, and once the AOA fell below a reset threshold, MCAS would move the stabilizer ANU to the original position and reset the system. At any time, the stabilizer inputs could be stopped or reversed by the pilots using their yoke-mounted electric stabilizer trim switches, which also reset the system after a 5 second delay.

The latter behavior is based on the assumption that flight crews use the trim switches to completely return the aircraft to neutral trim. In the FCC software version current at the time of the accident, if the original elevated AOA condition persists for more than 5 seconds following an MCAS flight control law reset, the MCAS flight control law will command another stabilizer nose down trim input (with the magnitude based on the AOA and Mach sensed at that time).

On all Boeing 737 models, column cutout switches interrupt stabilizer commands, either from the auto-flight system (e.g. FCC) or the electric trim switches in a direction opposite to elevator command. On the Boeing 737NG and Boeing 737 MAX, two column cutout switching modules, one for each control column, are actuated when the control columns are pushed or pulled away from zero (hands off) column position. When actuated, the column cutout switching modules interrupt the electrical signals to the stabilizer trim motor that are in opposition to the elevator command.

The MCAS function requires the stabilizer to move nose down in opposition to the column commands when approaching high angles of attack. To accommodate MCAS, the column cutout function in the first officer's switching module was modified to inhibit the aft column cutout switch while MCAS is active, allowing aircraft nose-down (AND) stabilizer motion with aircraft nose-up (ANU) column input. Once MCAS is no longer active, the normal column cutout function in the stabilizer nose down direction is re-instated.

Autopilot System and Flight Director (F/D)

(F/D) Autopilot Engagement Criteria

The Boeing 737-8 (MAX) Auto Flight Director System (AFDS) is a dual system consisting of two individual FCC and a single mode control panel.

The autopilot (A/P) functionality is provided by the FCC. The A/P for captain side is controlled by the FCC A and the A/P for first officer side is controlled by the FCC B.

Each A/P can be engaged by pushing a separate CMD (command) engage switch. A/P engagement is inhibited unless the following flight crew–controlled conditions are met:

- No force or very low force is being applied to the control wheel
- The STAB TRIM cut out switches is at NORMAL.
- Neither pilot main electric trim switches is activated.
- The Mode Control Panel (MCP) disengage bar is not activated.

Except during approach, only one A/P can be engaged at a given time.

- Pushing either A/P disengage switch,
- Column or wheel force override,
- Pushing either Takeoff/Go-around (TO/GA) switch:
- Pushing an illuminated A/P ENGAGE switch,
- Pushing the A/P DISENGAGE bar down,
- Activating either flight crew control main electric trim switch,
- Moving either STAB TRIM cutout switch to CUTOUT.

Autopilot Disengagement

During single channel operation, A/P automatically disengages when any of the following crew-controlled conditions occurs:

- With flaps not up or,
- Glide Slope (G/S) engaged,

Flight Director (F/D)

Flight Director (F/D) provides aircraft attitude guidance requires for the flight crew to obtain selected targets, such as altitude, speed, heading etc. The F/D functionality is provided by the FCC.

Left and right F/D switches activate command bars on the captain and first officer primary flight displays respectively.

The F/D takeoff mode is engaged by pushing the Takeoff/Go around (TO/GA) switches on either thrust lever. The Flight Mode Annunciation (FMA) displays F/D as the A/P status, TO/GA as the pitch mode, and HDG SEL as the roll mode.

The amber Pitch Limit Indicator (PLI) indicates the attitude at which stick shaker activation will occur for the existing flight conditions. The PLI is a function of the SMYD which uses AOA as an input.

Auto-throttle (A/T) System

The A/T system provides automatic thrust control from the start of takeoff through climb, cruise, descent, approach and go–around or landing.

The auto-throttle (A/T) system controls engine thrust in response to the mode selected by the flight crew through the Digital Flight Control System (DFCS), Mode Control Panel (MCP), Flight Management Computer (FMC) and ADIRU. The speed information taken from the ADIRU is used to calculate throttle lever rate commands to set engine thrust during changing flight conditions. All the information is processed by FCC A, which provide commands to the thrust lever servo motors controlling thrust lever movement

The flight crew may move the A/T switch to ARM. Arming the A/T is preparing the system to engage either on the N1, MCP SPD, or FMC SPD mode.

For each flight phase the flight crew can select the A/T N1 or speed modes from the MCP or directed by the FMC.

During takeoff, pushing TO/GA switch engages the A/T in N1 mode and causes the engine thrust to increase to the takeoff (TO) N1.

Throttle Hold (THR HLD) Mode

The throttle hold mode is automatic and the A/T goes into THR HLD mode during the takeoff ground roll. In this mode, the A/T removes power to the auto-throttle Servo Motor (ASMs) to prevent the A/T from moving the thrust levers during the takeoff roll and initial climb out. The A/T uses two separate functions to remove power from the ASMs. One is a software function and the other is a hardware function. When both throttle-hold functions agree and remove power to the servos, the A/T mode shows THR HLD on the FMA.

THR HLD mode starts when airspeed is > 84 knots CAS, and ends when barometric altitude is more than 800 feet above field elevation and at least 10 seconds has elapsed since lift-off.

During take-off the A/T will remain engaged if both left and right altitude and airspeed data are valid from the ADIRU.

Stall Warning System

Natural stall warning (buffet) usually occurs at a speed prior to stall. In some configurations the margin between stall and natural stall warning is less than desired. Therefore, an artificial stall warning device, a stick shaker, is used to provide the required warning.

Each control column has an eccentric weight motor which can vibrate the column to alert the pilots before a stall develops. The system is armed in flight at all times. The system is deactivated on the ground, except during the ground test.

Two independent, identical SMYD computers determine when stall warning is required based upon:

- Alpha vane angle of attack outputs
- ADIRU outputs
- Anti–ice controls
- Wing configurations
- Air/ground sensing
- Thrust
- FMC outputs.

The AOA sensor is connected to the SMYD and provides the measured angle of the direction of airflow relative to the fuselage. If the AOA sensor detects an excessive angle of attack compared to the design characteristic of the Boeing 737-8 (MAX), the SMYD will activate the stick shaker to provide aural and tactile alert to the flight crew.

Two SMYD computers provides output for stall warning to include stick shaker, Pitch Limit Indicator, and maneuver and operating airspeed limit.

The No.1 SMYD activates the Captain stick shaker, and the No. 2 activates the F/O stick shaker. Vibrations from either stick shaker can be felt in both columns through the mechanical column interconnect.

AUTO BRAKE DISARM Amber Light

The AUTO BRAKE DISARM amber light shows that there is a disarm condition in the autobrake system.

The AUTO BRAKE DISARM amber light comes on when the flight crew selects autobrakes and any of these conditions occur:

- • Malfunction in the autobrake system
- • Malfunction in the antiskid system
- • Autobrake system is manually disarmed.

To reset the auto brake disarm relay, move the AUTO BRAKE select switch to the OFF position. This will turn the amber AUTO BRAKE DISARM light off when the autobrake system is disarmed.

For takeoff, the AUTO BRAKE select switch is set to RTO. If the aircraft touches down with the switch still in the RTO position, the autobrake system will disarm and the AUTO BRAKE DISARM light will illuminate.

Take Off Configuration Warning Light
Takeoff configuration warning is armed when the aircraft is on the ground and either forward thrust lever is advanced for takeoff. The Takeoff configuration warning activates if the aircraft is not correctly configured for takeoff. One such condition is that that the leading edge devices are not in the normal takeoff position.

An intermittent warning horn sounds and the TAKEOFF CONFIG warning light illuminates when the takeoff configuration warning activates.

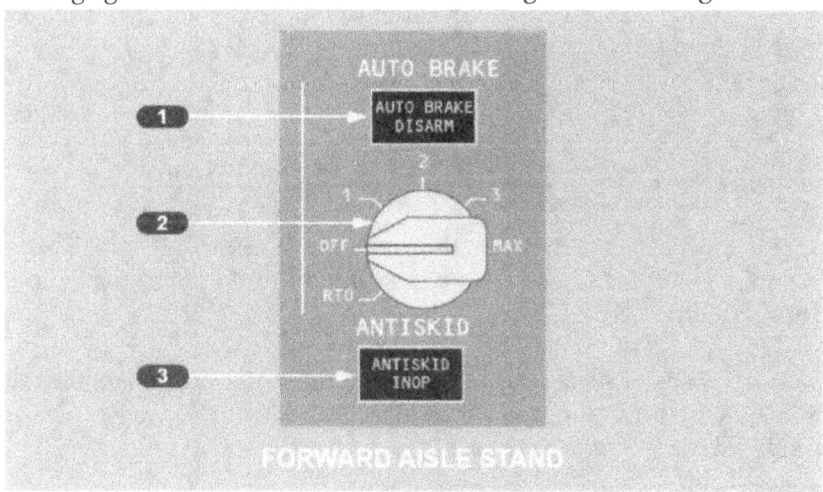

Figure 4: Auto Brake Selector and Auto Brake Disarm Light

Enhanced Ground Proximity Warning System (EGPWS): Bank Angle Alert
The Enhanced Ground Proximity Warning System (EGPWS) provides the aural alert BANK ANGLE, BANK ANGLE when excessive roll of the aircraft occurs.

The alert is based on radio altitude and bank angle:
- • From 5 feet to 30 feet AGL, the alert sounds when the bank angle exceeds 10 degrees
- • From 30 feet to 130 feet AGL, when the alert sound varies linearly from a bank angle of 10 degrees at 30 feet AGL, to a bank angle of 35 degrees at 130 feet AGL
- • Above 130 feet AGL, the alert sounds when the bank angle exceeds 35 degrees.

Air Traffic Controller (ATC) Transponder

The Boeing 737-8 (MAX) is equipped with two ATC transponders controlled by a single control panel. The aircraft ATC transponder system transmits a coded radio signal when interrogated by an ATC ground station. Altitude reporting capability is provided.

The aircraft ATC transponder replies to ATC ground station as follows:
- • The Mode A reply contains the aircraft identity code.
- • The Mode C reply contains the aircraft altitude.
- • The mode S replies contain a unique 24-bit aircraft address, identity code, altitude information, and Traffic Alert and Collision Avoidance System (TCAS) information.

The ALT SOURCE (altitude source) switch has two selection positions. The selection 1 enables altitude reporting from Air Data Computer (ADC) number 1 while the selection 2 is from the ADC number 2

Transponders may also transmit Automatic Dependent Surveillance-Broadcast (ADS-B) data which is downlinked to ATC and can be used for aircraft tracking.

The ATC radar system receives aircraft altitude data transmitted by the aircraft's selected transponder mode and the ATC radar system displayed the ground speed based on the aircraft movement calculation on the radar system.

Boeing 737-8 (MAX) Fault Handling System

The Boeing 737-8 (MAX) is equipped with the Maintenance Awareness System (MAS) that provides the information directly related to airworthiness and system information. The system includes the MAINT light (amber color) on the overhead panel, stored fault information which can be accessed from the Multi-Function Display (MFD) control on the pedestal.

To conduct troubleshooting, can use the Interactive Fault Isolation Manual (IFIM) provides a searchable database to quickly isolate the cause of each aircraft fault.

The IFIM searchable pages show the IFIM task which guided the engineer to conduct the fault isolation and the engineer should perform the rectification by following the tasks which may include:

The IFIM task or AMM task may include the procedure for testing the system after rectification of the fault. After rectification, if the associated OMF maintenance message is found to be NOT ACTIVE, the problem is considered to be solved, and if the associated OMF maintenance message is ACTIVE, the problem still exists.

In some circumstances, operator may elect to defer maintenance. The situations in which it is permissible to dispatch an aircraft with an existing fault and the necessary modifications to maintenance and operational procedures are found in the Minimum Equipment List (MEL). If allowed by the MEL, the engineer may dispatch the aircraft with the fault status still in ACTIVE.

IAS Flag Fault Isolation and Rectification Procedure

The IAS Flag is an airspeed indicator fault that was displayed in the PFD. In the chapter Observed Fault of the FRM referred the IAS Flag as "airspeed display is blank" or "SPD flag."

If the flight crew did not enter the fault code or only mention the fault description in the aircraft flight and maintenance log, the engineer should search the symptom on the OMF that will shows the related faulty component and its maintenance message. Refer to the maintenance message from the OMF, the engineer search in the IFIM that will show the IFIM task number.

Without the OMF maintenance message, the engineer may search the symptom in the IFIM.
)

If the engineer directly enters the fault code of 342 612 31 (for the "captain airspeed is blank"), the IFIM will directly shows the IFIM task number 31-63-00-810-803 (Airspeed Display Is Blank (Captain's) - Fault Isolation). If the engineer directly enters the fault code of 341 301 31 (for

"SPD flag" shows on PFD), the IFIM will shows the IFIM task number 31-98-00-810-803 which is longer than IFIM task 31-63-00-810-803 and requested engineer to do more in the OMF menu to find the correlated maintenance message.

The IFIM task number 31-63-00-810-803 directs the engineer to:

- Conduct a self-test of the ADIRU and correct any resulting faults.
- Check in the OMF for related faults and correct any found.

ALT Flag Fault Isolation and Rectification Procedure

The ALT Flag is an altitude indicator fault that was displayed in the PFD. If the engineer directly enters the fault code of 342 302 31 the IFIM will directly shows the IFIM task number 31-98-00-810-803 which directs the engineer to check in the OMF for related faults, and correct any found.
The rectification of the problem ALT flag is depending on the maintenance message as shown on the OMF.

IAS Disagree Fault Isolation and Rectification Procedure

The IAS Disagree was displayed in the PFD if there is disagreement of speed indicator between the captain and first officer display. In the chapter Observed Fault of the FRM the fault code of IAS Disagree is define as SPEED DISAGREE or SPD DISAGREE.

Without the OMF maintenance message, the engineer may search the symptom in the IFIM.

If the engineer directly enters the fault code of 341 302 31 the IFIM will directly show the IFIM task number 34-10-00-810-801 which directs the engineer.

The replacement of the AOA as mentioned above related to the visual inspection of the AOA unit. In this IFIM task did not mention the measurement of AOA value in the SMYD.

ALT Disagree Fault Isolation and Rectification Procedure

The ALT Disagree was displayed in the PFD if there is disagreement of altitude indicator between the captain and first officer display. In the chapter Observed Fault of the FRM the fault code of ALT Disagree is define as ALT DISAGREE.

Selecting the IFIM task 34-20-00-810-801 describe the fault isolation for ALT DISAGREE Shows on PFD (Captain's) which further directs the engineer. to

The replacement of the AOA as mentioned above related to the visual inspection of the AOA unit. In this IFIM task did not mention the measurement of AOA value in the SMYD.

Angle of Attack failure

According to the Boeing 737-8 (MAX) Illustrated Part Catalogue (IPC), the installed Angle of Attack sensor part number is 0861FL1. Refer to the Fault Reporting Manual (FRM).

Entering those fault codes into IFIM will lead to IFIM task 31-98-00-810-803 (Standard Procedure for Observe Fault with Correlated Maintenance Message). The IFIM task contain the correlated OMF maintenance message 34-21123 (AOA signal is out of range) which lead to the IFIM task 34-21-00-810-828 (Angle of Attack Signal out of Range).

The initial evaluation of the IFIM task 34-21-00-810-828 requires the test of AOA sensor referring to the AMM TASK 34-21-05-400-801 which contains measurement of the AOA vane angle value via SMYD.

AOA Disagree Fault Isolation and Rectification Procedure

AOA DISAGREE message will appear in amber color on the captain and first officer PFD if the left and right AOA sensor values disagree by more than 10 degrees for more than 10 seconds.
The FRM stated that the AOA DISAGREE fault is an observed fault where the problem symptoms can be sensed by the flight crew or the maintenance crew. The fault codes of AOA DISAGREE are defined in the FRM.

Entering the fault code of 342 303 31 or 342 303 32 for AOA DISAGREE into the IFIM will show the IFIM task number 34-20-00-810-803 or 34-20-00-810-804 with the task title of AOA DISAGREE shows on captain or first officer PFD respectively. The detail task is check in the OMF for related faults and correct any found.

If the flight crew did not enter the fault code and only mention AOA DISAGREE, the engineer should search the symptom on the OMF which might show any maintenance message related with AOA DISAGREE.

Selecting the two faults on the top row of the list will show the fault code 342 303 31 and 342 303 32 which consistent with the fault code stated in the FRM.

Selecting the maintenance messages will shows the maintenance message containing text "AOA" and/or "DISAGREE". From the 47 maintenance messages, there were 18 maintenance messages related with the AOA.

Meteorological Information

The *Badan Meteorologi Klimatologi dan Geofisika* (BMKG – Bureau of Meteorology, Climatology and Geophysics) provided enhanced infrared satellite images. The enhanced infrared satellite images at 2320 UTC (0620 LT) up to 2330 UTC (0630 LT) indicated that the cloud top temperature at the accident site (red circle) was from 0 up to 8°C.

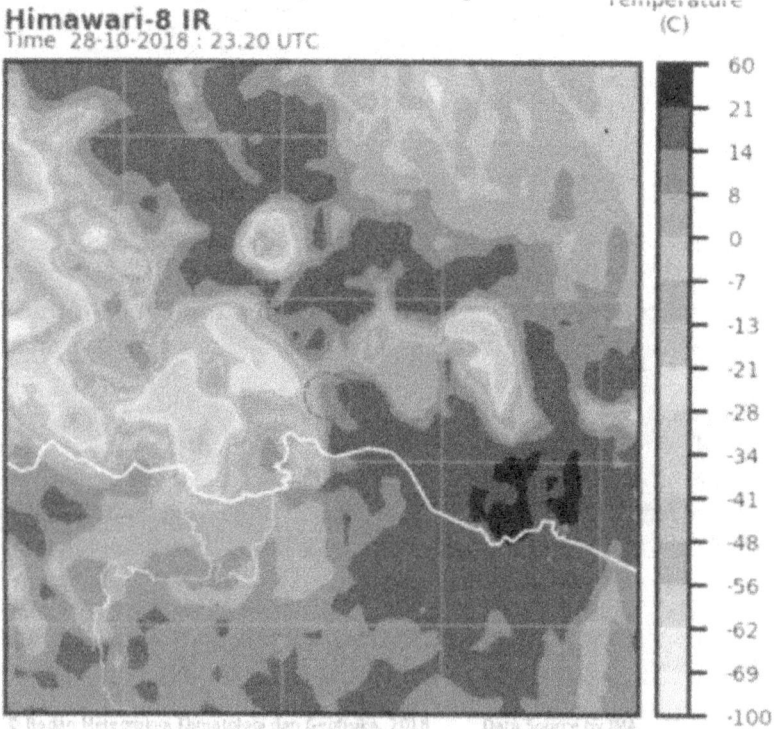

Figure 5: Enhanced infrared satellite image at 2320 UTC (0620 LT)

Aids to Navigation

ABASA 1C
The runway 25L utilized RNAV-1 Standard Instrument Departure (SID), the SID of ABASA 1C requires after departure to climb on heading 248°, at or above 3,000 feet, and then turn left direct to BUNGA – RATIH – LARAS – TOMBO – ABASA..

Automatic Dependent Surveillance – Broadcast

Automatic Dependent Surveillance – Broadcast (ADS–B) is a surveillance technology in which an aircraft determines its position via satellite navigation and periodically broadcasts it, enabling it to be tracked.

The "automatic" in the ADS-B means that the technology does not require flight crew or external input. The "dependent" means its surveillance process depends on data on-board aircraft systems to provide surveillance information to the receiver and "broadcast" means the originating source has no knowledge of who receives the data and there is no interrogation or two-way contract.

Several receivers have been installed in several places including in the Jakarta Air Traffic Services Center (JATSC). The PK-LQP aircraft had ADS-B capability and the investigation retrieved the aircraft broadcasted data from the JATSC facility.

The flight track of the LNI610 based on the ADS-B data superimposed on Google Earth.

Figure 6: The flight track of LNI610 based on ADS-B

Communications

All communications between Jakarta air traffic controllers and the flight crew were recorded by ground based automatic voice recording equipment and the cockpit voice recorder. The quality of the aircraft recorded transmissions on the ground based automatic voice recording equipment was good.

CHAPTER 2

FLIGHT RECORDERS

Digital Flight Data Recorder (DFDR)

The aircraft was fitted with a FA2100 DFDR manufactured by L3 Technologies with part number 2100-4945-22 and serial number 001261573.

On 1 November 2018, the Crash Survivable Memory Unit (CSMU) of the DFDR was recovered by the search team. The CSMU was transported to the KNKT recorder facility for data downloading. The read-out was performed by KNKT investigators with the participation of the Australian Transport Safety Bureau (ATSB), Transport Safety Investigation Bureau (TSIB) of Singapore and National Transportation Safety Board (NTSB) of United States of America as Accredited Representatives and Boeing.

The memory unit recorded 1,790 parameters and approximately 69 hours of aircraft operation, which contained 19 flights including the accident flight.

The DFDR parameters graphs of the accident flight showed:

> 1. Since the beginning of the plot, the DFDR recorded differences between the left and right AOA. The AOA data considered invalid when the aircraft was not moving. The AOA differences between left and right were constant, about 21°, from

the time the aircraft accelerated for take-off until the end of recording.

2. During rotation, the aircraft lifted off then briefly touched prior to becoming fully airborne.

3. There were no recorded indications of autopilot engagement during the flight.

4. The computed airspeed and altitude parameters showed differences (split) between left and right from the time the aircraft became airborne until the end of recording.

5. For most of the flight, the recorded altitude fluctuated around 5,000 feet and never maintained a constant altitude.

6. The left control column stick shaker activated just after the aircraft became airborne. It temporarily stopped about 2322 UTC when the aircraft descended with flaps extended. About 15 seconds later the left control column stick shaker activated again and was continuously active until the end of recording.

7. About 2322 UTC the flaps were selected to zero and few seconds later, the automatic trim down command triggered by MCAS became active. The automatic trim down command stopped when the trim manual up activated.

8. About 2325 UTC, the flaps were selected to zero and remained at zero until the end of recording. The automatic trim down activated repeatedly until the end of the recording, and was generally followed by manual electric trim up. From the time the flaps were set to zero until the end of the recording there were at least 26 automatic trim down commands and at least 34 manual electric trim up inputs.

9. During the activation of the manual electric trim up and automatic trim down, corresponding changes in the pitch trim position were recorded. For most of the accident flight, the pitch trim position was above 5 units after the activations of manual electric trim up and decreased after the activation of automatic trim down.

10. From about 2331 UTC until the end of the recording, the activations of manual electric trim up were shorter than the activations of automatic trim down and the pitch trim position gradually decreased to near zero at the end of recording. When the pitch trim position was about 1.5 units, the aircraft started to

descend rapidly and the air speed and engine power (N1) increased.

11. At 23:31:54 UTC, the DFDR stopped recording.

Cockpit Voice Recorder

The aircraft was fitted with a FA2100 CVR manufactured by L3 Technologies with part number 2100-1925-22 and serial number 001257879.

On 14 January 2019, the Crash Survivable Memory Unit (CSMU) of the CVR was recovered. The Underwater Locator Beacon (ULB) had detached from the CVR CSMU and was not recovered. The CVR CSMU was transported to the KNKT recorder facility for data downloading. The read-out was performed by KNKT recorder specialist with the participation of the National Transportation Safety Board (NTSB) of United States of America and Transport Safety Investigation Bureau (TSIB) of Singapore.

The memory unit recorded good quality; recorded data on four channels and was 2 hours and 4 minutes in duration. The data recorded on the CVR included the accident flight preparation and was continuous until the end of the accident flight.

The relevant excerpt of CVR was as follows:

Time (UTC)	Event
22:18:48	The FO advised the Captain that this flight was not his actual schedule. The FO was called at 4 o'clock in the morning and informed the revision of the original schedule.
22:25:39	The Captain advised the FO that he was having flu. The CVR recorded the Captain coughed about 15 times within an hour during the preflight.
22:47:59	Engineer came to cockpit and advised the Captain that he would be on board the aircraft to Pangkal Pinang. The engineer explained that he had not rated for Boeing 737-8 (MAX) and not authorized to release the aircraft.
23:08:02	Flight attendant advised the pilots that total passengers on board were 181. The Captain then advised to the FO that the person on board were 181 plus 7.
23:09:12	The FO advised the Ground controller that the POB was 188 and requested pushback clearance. The Ground controller asked the LNI610 to standby.
23:10:15	The Ground controller issued clearance for push back
23:15:57	The Ground controller issued clearance for taxi
23:19:31	The Jakarta Tower controller issued takeoff clearance using runway 25L. The FO readback the clearance.
23:20:32	The Enhanced Ground Proximity Warning System (EGPWS) sounded "V1".
23:20:33	The FO called "rotate".
23:20:41	The Captain exclaimed about what happened to the aircraft.
23:20:44	The FO advised to the Captain "auto brake disarmed".
23:20:47	The FO advised the Captain "indicated airspeed disagree".
23:20:49	The FO advised the Captain "positive rate" and was responded "gear up".

23:20:52	The FO exclaimed about what happen with the aircraft and asked the Captain whether the Captain wanted to return. The Captain did not respond to the FO.
23:21:03	The Jakarta Tower controller instructed the LNI610 to contact director (Terminal East (TE) controller) and the instruction was acknowledged by the FO.
23:21:18	The FO advised to the Captain "Altitude Disagree" and was acknowledged.
23:21:22	The FO made initial contact with the TE controller and responded that the aircraft was identified on the radar screen. Thereafter, the TE controller instructed the LNI610 to climb to altitude 27,000 feet.
23:21:28	The FO asked to the TE controller of the aircraft altitude as shown on the TE controller radar display. The TE controller responded that the aircraft altitude was 900 feet and was acknowledged by the FO.
23:21:37	The Captain instructed the FO to perform memory item for airspeed unreliable.
23:21:44	The FO requested Captain intention of the altitude to be requested to the TE controller.
23:21:45	The Captain responded: "Yeah…request uh…proceed". The FO then suggested to left downwind who then the Captain instructed the FO to request to some holding point.
23:21:52	The FO requested approval to the TE controller to proceed to any holding point.
23:22:01	The FO suggested whether the Captain need flap who then affirmed.
23:22:03	The FO suggested flap 1 to the Captain who then affirmed.
23:22:04	The TE controller asked the LNI610 of the problem and FO responded that they experiencing flight control problem.
23:22:12	The FO advised to the Captain "Feel Differential".

23:22:14	The Captain handed over the control to the FO who then responded to standby. The Captain requested that the FO over the control. The FO responded to standby
23:22:21	The TE controller advised to the LNI610 that the aircraft descended to "ONE SEVEN HUNDRED" and the TE controller asked the intended altitude.
23:22:24	The FO suggested the Captain "you want flap up?" who then affirmed by the Captain.
23:22:28	The FO suggested to the Captain the altitude of 6,000 feet. The Captain requested to 5,000 feet.
23:22:30	The FO advised the TE controller that the intended altitude was 5,000 feet.
23:22:31	The TE controller instructed the LNI610 to climb and maintain altitude of 5,000 feet and to turn left heading 050°.
23:22:32	EGPWS sounded: "BANK ANGLE BANK ANGLE".
23:22:35	The FO read back the TE controller instruction.
23:22:41	The Captain "Ok flap 1".
23:22:44	The FO "5,000".
23:22:57	The FO asked the TE controller the speed as indicated on the radar display.
23:23:00	The EGPWS sounded "AIR SPEED LOW AIR SPEED LOW". The TE controller responded that the speed shown on the ASD was 322 knots of ground speed.
23:23:09	The Captain commanded "Memory item, memory item".
23:23:17	The FO: "Feel differential already done, auto brake...engine start switches off, what's the memory item here". The Captain responded "check".
23:23:34	The FO: "flight control" who then responded "yeah" by Captain.
23:23:39	Similar sound of turned paper pages.
23:23:48	The FO: "flight control low pressure..."
23:23:52	Sound of altitude alert tone.
23:23:53	The FO: "...nine point eight".

Time	Event
23:24:03	Sound of trim wheel movement.
23:24:05	The FO: "Feel differential pressure".
23:24:09	The FO: "Which one …. [Unintelligible]".
23:24:11	The Captain: "no…no…air speed unreliable, air speed unreliable". The FO then responded "sorry".
23:24:19	The FO: "airspeed unreliable, standby".
23:24:31	The FO: "where is the…"
23:24:34	The FO: "no airspeed"
23:24:43	Altitude alert tone
23:24:46	The FO: "airspeed, airspeed"
23:24:53	Similar sound of turned paper pages.
23:24:57	The TE controller instructed the LNI610 to turn left heading 350° and maintain altitude of 5,000 feet. The FO read back the instruction.
23:25:11	The FO: "heading 350…there is no airspeed unreliable".
23:25:17	The FO advised the Captain "TEN point ONE". Note: 10.1 is the page number of Airspeed Unreliable NNC
23:25:20	The FO read Airspeed Unreliable checklist: "Condition: Airspeed or Mach indications are suspected to be unreliable. To identify a reliable airspeed indication, if possible, to continue the flight using the Flight with Unreliable Airspeed. Autopilot if engaged disengages … Already. Auto throttle if engaged disengage".
23:25:41	The Captain: "disengaged". The FO continued reading Airspeed Unreliable checklist: "FD switches both off".
23:25:43	The Captain: "Off"
23:25:44	Sound of altitude alert tone.
23:25:46	The FO continued reading Airspeed Unreliable checklist: "Set the following gear up pitch attitude and thrust. Flaps up four degrees and seventy five percent N1".
23:26:00	Similar sound of trim wheel movement.
23:26:02	The Captain: "Yeah"

Time	Event
23:26:04	The FO: "75%".
	The Captain: "yeah".
23:26:11	The FO: "already now"
	The Captain:" "yeah".
23:26:15	The FO continued reading Airspeed Unreliable checklist: "the following are reliable attitude, N1, ground speed, or radio altitude"
	The Captain: "check".
23:26:28	The FO continued reading Airspeed Unreliable checklist:
	"Note: Stick shaker, over speed warning and AIRSPEED LOW alerts may sound erroneously"
23:26:32	The TE controller instructed the LNI610 to turn right heading 050° and maintain altitude of 5,000 feet. The FO read back the instruction.
23:26:59	The TE controller instructed the LNI610 to turn right heading 070° to avoid traffic.
	The FO was continuing to read the Airspeed Unreliable checklist: "The Flight path vector and pitch altitude indicator may be unreliable, refer to the Flight with Unreliable Airspeed table".
23:27:07	The TE controller called the LNI610 with no respond and repeated the call 5 seconds after
23:27:13	The FO responded "go ahead"
23:27:15	The TE controller instructed the LNI610 to turn right heading 090°. The FO read back the instruction.
23:27:23	The FO continued reading the Airspeed Unreliable checklist: "Performance Inflight chapter and set the pitch attitude and thrust setting for the current aircraft"
23:27:28	The TE controller revised the instruction to stop the turn and fly heading 070°. The FO read back the instruction.
23:27:37	The FO continued reading the Airspeed Unreliable checklist: "When in trim and stabilized, cross check the captain, first

	first officer and standby airspeed indicator". The Captain then responded "check".
23:27:43	The FO: "280, 280"
23:27:46	The FO continued reading the Airspeed Unreliable checklist: "That differs more…differs by more than 20 knots or 0.03 Mach from the airspeed shown in the table should be considered unreliable". The Captain then responded "yeah".
23:27:58	The FO suggested to the Captain that he would check the performance inflight chapter who then responded "yeah".
23:28:03	Sound similar to interphone two tone chime
23:28:05	Sound similar to FA Captain picked up the handset
23:28:09	The FA: "Yes sir … hello" and the FO instructed the FA to come in the cockpit.
23:28:12	Sound similar to FA handset being stowed.
23:28:15	• Sound similar to flight deck door opened. • The TE controller provided traffic information of an Airbus A320 position at 10 o'clock 11 Nm leaving 5,000 descending to 2,000 south east bound.
23:28:18	• Sound similar to thump • The Captain requested the FA to call the engineer to the cockpit
23:28:19	The FO repeated the Captain request to call the engineer
23:28:25	The FO responded the traffic information from the TE controller was monitored.
23:28:36	Sound similar to handset of FA being picked up.
23:28:37	Sound similar to flight attendant call, two single chimes.
23:28:39	Sound similar to handset of FA being stowed
23:28:41	Sound similar to cockpit door being opened.

23:28:43		The TE controller instructed the LNI610 to turn left heading 050° and maintain an altitude of 5,000 feet. The FO read back the instruction: "turn left ZERO FIVE ZERO FIVE THOUSAND Lion SIX ONE ZERO"
23:28:54		Sound of FA station handset calling FA other station.
23:28:55		Sound of two chimes twice.
23:28:55		The Captain asked someone to look what happened
23:28:58		FA communicated with other FA explained that there was technical issue in the cockpit, and asked the other FA to review, and was acknowledged by the other FA.
23:29:14		The FO advised: "Ok we are already gear up 5,000…"
23:29:18		Sound similar to altitude alert.
23:29:19		The FO: "…with".
23:29:28		The FO: "Fly up".
23:29:31		The FO: "Air cond…off schedule descend understood"
23:29:37		The TE controller questioned the LNI610 whether the aircraft was descending.
23:29:40		The Captain: "we have some problem…"
23:29:41		The FO advised the TE controller that they had flight control problem and flew the aircraft manually. The TE controller responded by instructing to maintain the aircraft heading to 050 and to contact the Arrival (ARR) controller on frequency 125.45.
23:29:50		Communication heard from ground Lion Air frequency.
23:30:02		The FO contacted the ARR controller and advised that they were experiencing flight control problem. The ARR controller advised LNI610 to prepare for landing on runway 25 Left and instructed them to fly heading 070°. The instruction was read back by the FO.
23:30:24		The FO: "there is 1,7" who then the Captain responded: "yeah".
23:30:29		The FO: "62.3, so our weight is 62.3, so for 5,000 we have speed".

23:30:48	The Captain asked the FO to take over the aircraft control for a while.
23:30:53	Sound of altitude alert tone.
23:30:54	The FO replied that he took over the aircraft control.
23:30:57	The Captain requested to the ARR controller using call sign LNI650 to direct to waypoint ESALA due to weather and was approved and instructed to maintain five thousand. The Captain read back the instruction
23:31:07	The FO "Wah, is very "
23:31:08	The Captain advised the ARR controller that the altitude of the aircraft could not be determined due to all aircraft instruments indicating different altitudes. The Captain used the call sign of LNI650 during the communication. The ARR controller acknowledged then stated "LNI610 no restriction".
23:31:22	The LNI610 Captain requested the ARR controller to block altitude 3,000 feet above and below for traffic avoidance. The ARR controller asked what altitude the flight crew wanted.
23:31:27	Sound of altitude alert tone
23:31:33	The FO exclaimed that the aircraft was flying down.
23:31:34	The Captain responded the ARR controller question: "FIVE THOU..".
23:31:36	The FO exclaimed the aircraft was flying down who then the Captain responded: "it's ok". The ARR controller approved the LNI610 request.
23:31:39	Sound of single chime.
23:31:41	Sound of over speed warning.
23:31:42	The FO: "fly up".
23.31.50	Sound of single chime interphone tone.
23.31.51	EGPWS: "TERRAIN, TERRAIN".
23:31:53	EGPWS: "SINK RATE".
23:31:55	The CVR stopped recording.

CHAPTER 3

ADDITIONAL INFORMATION

Wreckage and Impact Information

The search team operated in conjunction with the Indonesia Search and Rescue team with the mission to recover remaining of the victims. The search team utilized a Remotely Operated Vehicle (ROV) equipped with an under-water camera, side scan sonar and 4 Under-water Locator Beacon (ULB) locators.

The search team identified the wreckage on the floor of the Java Sea near Tanjung Karawang, about 32 Nm from Jakarta on bearing 056°. The wreckage was scattered over an area about 200 by 140 meters which was about 370 meters from the last aircraft position recorded on the ADS-B (Automatic Dependent Surveillance – Broadcast).

The FDR CSMU was found at 5°48'43.20" S 107°7'37.60" E and the CVR CSMU was found at 5°48'46.52" S 107°7'36.85" E, which were within the wreckage distribution area. Several parts of the aircraft were recovered and transported to Jakarta International Container Terminal (JICT). The wreckage recovered was from all sections of the aircraft from the forward to the aft.

The recovered wreckage identified consisted of, but was not limited to, parts from the left and right engines, both main landing gears, parts of the empennage, parts of the forward and aft left passenger doors, parts of the wings, a flight crew oxygen bottle, and parts of a circuit breaker panel. The damage to the aircraft suggested a high energy impact.

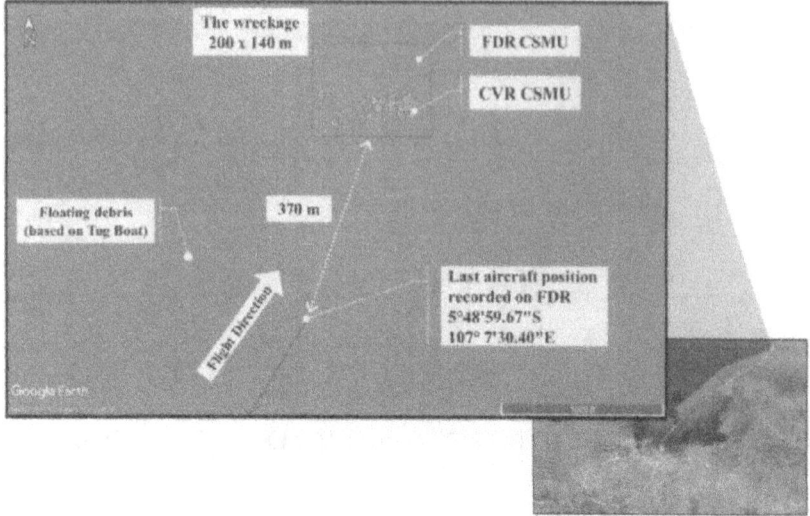

Figure 7: The wreckage distribution

Medicaland Pathological Information

Not relevant to this accident.

Fire

There was no evidence of in-flight fire.

Survival Aspects

PK-LQP was equipped with an airframe-mounted low frequency underwater locator beacon (ULB) which operated at a frequency of 8.8 kHz. The beacon is included in ICAO standards[12]. The purpose of the beacon is to aid in the location of submerged aircraft. During the search phase, multiple surveys were conducted to detect a signal at 8.8 kHz. No such signals were detected in the area where wreckage was recovered.

The beacon received ETSO-C200a approval. ETSO-C200a invokes minimum performance standard SAE document AS6254A Minimum Performance Standard for Low-Frequency Underwater Locating Devices (Acoustic) (Self-Powered), dated 6 December 2015, which includes requirements to operate to a depth of at least 20,000 feet, impact shock 5 milliseconds 1,000 G and static crush of 5,000 lbf.

The beacon was mounted on the forward side of the nose pressure bulkhead.

Boeing selected the location after many distinct configurations were evaluated based on safety and certification focused tests and analysis. Most of the preferred installation locations could not be used because they proved to be incompatible with EASA and FAA Non-Rechargeable Lithium Battery certification requirements or they did not meet the ICAO empennage and wings exclusion.

CHAPTER 4

TESTS AND RESEARCH

Installation Test AOA Sensor with Known Bias

On 15 November 2018, with approval from KNKT and under direction and supervision of the NTSB, Boeing and the NTSB conducted an installation test of an AOA sensor on an exemplar Boeing 737-7 (MAX) located at the Boeing Field Flight Line. This test was intended to demonstrate if the AMM installation test, task 34-21-05-400-801, was robust enough to ensure that a bias in an AOA sensor could be identified/detected using the installation and alternate test procedure.

1. The tests were conducted on a production 737-8 (MAX) aircraft in serviceable condition inside a hangar with adequate lighting.

2. The first test consisted of installing a known serviceable AOA sensor in the left position of the aircraft on the production aircraft.

3. The second test consisted of installing an AOA sensor with an induced or known bias of approximately 33° (modified to have 33° bias) in the left position of the aircraft.

4. The tests were done with and without entering good weight data into the FMC.

The tests conducted as follow:

1. Test of the serviceable AOA sensor

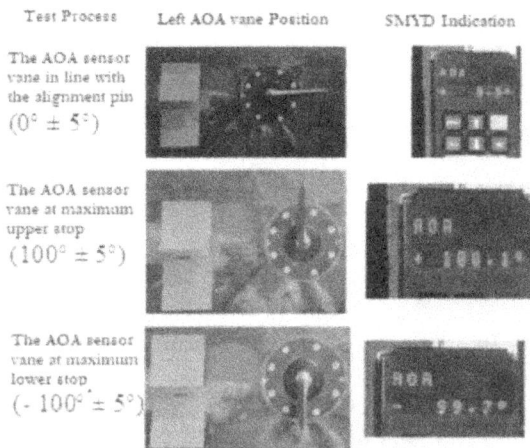

Figure 8. Test of the serviceable AOA sensor.

2. Test of the AOA sensor with known bias. The SMYD was set up to show the value of vane position angle measured by the left AOA sensor.

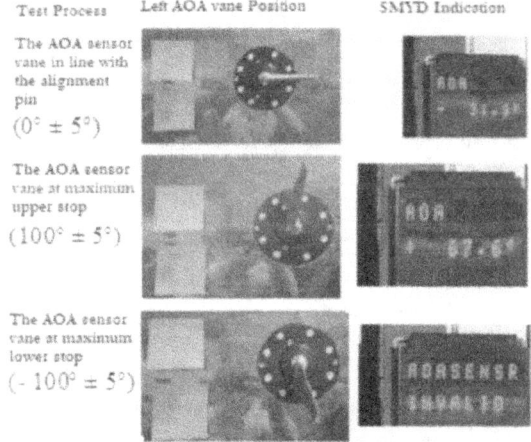

Figure 9. Test of the AOA sensor with known bias.

Conclusion:

1. With the serviceable (original) AOA sensor installed on the aircraft, the results of the alternative installation test indicated that the left AOA sensor met the AMM requirements.

2. With the biased AOA sensor, the test found that the vane angle values exceed the limits as follows:

a. When the vane was at its zero position, the SMYD displayed -31.9° (the misalignment angle) instead of 0° ± 5°

b. When the vane positioned at its maximum upper stop, the SMYD displayed +67.6° instead of +100° ± 5°.

c. When the vane positioned at its maximum lower stop, the SMYD displayed the text "AOA SENSR INVALID".

3. The alternative method of the installation test in the AMM will successfully detect a mis-calibrated AOA sensor.

Observation to the Xtra Aerospace

On 12 December 2018, the National Transportation Safety Board (NTSB), Federal Aviation Administration (FAA), Boeing, and Collins Aerospace representatives convened at Xtra Aerospace LLC (Xtra) facility in Miramar, Florida to review and document the maintenance records for the AOA Sensor, part number 0861FL1/serial number 14488 (AOA that was installed on the accident aircraft at the time of the accident) and to review the test equipment and procedures used during the repair process.

Collins Aerospace provided an exemplar AOA sensor obtained from the manufacturer's new stock, part number 0861FL1, serial number 22160, to support the investigation. The group observed an Xtra Aerospace technician perform the test procedures contained in CMM 34-12-34, Revision 8, on the exemplar unit. Note: CMM Revision 8 was in effect at the time S/N 14488, the accident AOA sensor, was repaired by Xtra.

An Xtra Aerospace technician performed tests 3.A through 3.E from the CMM Revision 8 Testing and Fault Isolation section. The tests performed were the Insulation Resistance test, Vane Friction test, Heater Current test, and Alignment Accuracy test. For this demonstration, the technician used a North Atlantic 8810A Angle Position Indicator (API) to measure the 0861FL1 resolver outputs. A Peak Electronics SRI-201B API ("Peak API") in "relative" mode, is used to read the synchro connected to the vane zero/indexing test stand (Note the Peak API could also be used for measuring the AOA resolver outputs, but for this demonstration the North Atlantic 8810A was used). The CMM specifies that resolver angles should be measured using a North Atlantic Model 8810A Angle Position Indicator (API), but includes the note "Equivalent substitutes may be used." The Peak SRI-201B is not listed in the CMM.

Xtra Aerospace utilized several pieces of test equipment to complete repair and evaluations on the AOA sensor that were not specified in the

CMM Revision 8. Xtra Aerospace instead utilized the following equipment in service at the time of repair of S/N 14488:

• Peak Electronics SRI-201B (Model 7724-00-2) (Peak API) (quantity 3); and

• North Atlantic 8810A (quantity 1).

According to the Repair Station Manual the level of accuracy should be equal or better than that recommended by the Original Equipment Manufacturer (OEM).

To meet the requirements of CMM Revision 8, Xtra Aerospace assessed the Peak API and North Atlantic 8810A for an equivalency justification. Xtra's engineering department prepared the equivalency report, which was approved by their quality assurance department. The FAA Flight Standards District Office (FSDO) accepted the test equipment equivalency report and the test equipment was considered by the FSDO to be in conformance with the CMM.

The Peak API has an additional mode of operation not available in the North Atlantic 8810, or described in the CMM test procedures. The additional mode of operation is "relative" mode. It was observed that Xtra Aerospace utilized a Peak API, set in relative mode, as a component of the vane indexing/zero fixture setup.

The Peak API has an additional mode of operation not available in the North Atlantic 8810, or described in the CMM test procedures. The additional mode of operation is "relative" mode. It was observed that Xtra Aerospace utilized a Peak API, set in relative mode, as a component of the vane indexing/zero fixture setup.

When the switch is set to the ABS (absolute) position, and the vane is locked in the vane zero position, the resolver outputs will read 45° on the Peak API (note: the resolvers are calibrated to output 45° with the vane at its zero position). If the switch is then moved to the REL (relative) position, the Peak API will display 0°, as a 45° offset had been established. Moving the AOA vane to a new position will result in the API displaying the actual resolver angle minus 45°. The -45° offset is constant throughout the full range of the vane rotation, so long as the REL/ABS switch remains in the REL position.

NTSB, FAA, Boeing and Collins Aerospace observed an Xtra Aerospace technician perform a vane-slinger-shaft assembly (VSS) removal and replacement (R&R) per CMM Revision 8 on the AOA sensor exemplar unit. During VSS R&R, the resolver gears and damper gear are initially disengaged from the main gear (which is fixed to the vane shaft) allowing the resolvers to rotate independently. Further, as part of VSS removal, the main gear is actually removed from the original VSS, insuring the resolvers are no longer retained in their original position as they can independently rotate to a new position.

After witnessing the vane-slinger-shaft replacement, a procedure was developed to determine if a 25-degree bias could inadvertently be introduced into both resolvers of an AOA 0861FL1 sensor if calibrated and tested using the Peak API in relative mode. 25° was an arbitrary non-zero angle chosen as a representative example.

The exemplar AOA unit was secured in the vane zero fixture, with both resolver gears disengaged from the main gear but positioned to 45°. The REL/ABS switch on the Peak API was positioned to ABS. Resolver 2 was then rotated until the display on the Peak API displayed 25° (Resolver 1 was not adjusted; its position remained at 45°). With the Peak API connected to Resolver 2, the REL/ABS switch was moved to the REL position at which time the Peak API display changed to indicate approximately 0°

With the REL/ABS switch still in the REL position, the resolver gears were engaged with the main gear, and the calibration procedure was performed per the CMM Revision 8. The resolvers were independently calibrated by rotating each resolver until the display of Peak API displayed an output of approximately 45°.

After the resolvers were calibrated to 45° with the REL/ABS switch in the REL position, the AOA outputs were checked at the CW and CCW rotational stops. The AOA outputs were found to be approximately ± 100 degrees per the design. The REL/ABS switch was then positioned to ABS. The resulting vane angle was recorded with the vane positioned in the vane zero fixture and at the CW/CCW mechanical stops. The angles recorded for Resolver 1 and Resolver 2 resulted in a 25-degree bias over the full range of vane travel. The test demonstrated that an AOA sensor calibrated and tested with a Peak API in relative mode could result in an equal bias introduced into both resolvers. The bias would not be detected during either AOA sensor calibration or CMM Revision 8 return-to-service testing.

In February 2019, Collins Aerospace repeated the Peak API offset demonstration at its facility. The test was repeated for the benefit of

NTSB and FAA personnel who did not witness the original demonstration at Xtra Aerospace. The procedures followed during the February 2019 demonstration were fundamentally the same as those performed at Xtra Aerospace in December 2018. The conclusions were identical. First, that an equal offset could inadvertently be introduced to both resolvers. Second, that the magnitude of the offset is essentially random. And third, that the offset could go undetected through the CMM return-to-service tests.

These test results suggest that there was a possibility of differences or a bias if the REL/ABS toggle switch was inadvertently selected to REL position.

Engineering Simulator

On 3-6 December 2018, the investigation team conducted aircraft simulator exercises in the Boeing engineering simulator configured for Boeing 737-8 (MAX). The team consisted of representatives from KNKT, NTSB, FAA, Boeing, Indonesia DGCA and Lion Air.

The main objective of the simulator exercise was to provide a better understanding regarding the accident and previous flights, which included the following activities:

1. Documented the various messages and lights:
a. IAS Disagree

b. ALT Disagree

c. Mach Trim Fail

d. Speed Trim Fail

e. Feel Differential Pressure Light

2. Documented the following events:
a. Flaps 5 to Flaps Up

b. Split AOA

c. MCAS activation durations

d. Runaway Stabilizer Trim

3. Documented the various failure modes and flight deck effects associated with those failures

4. Documented procedures and crew response including prioritization of multiple messages

5. Understanding the flight crew workload during different scenarios.

The simulator setup was as follows:

1. The airport was Jakarta, runway 25L for departure.

2. Left seat/Captain is Pilot Flying (PF), Right seat/First Officer is Pilot Monitoring (PM) (unless otherwise noted).

3. Weather was similar to the accident flight, which were wind 3 knots from 160°, visibility 8 km, clouds scattered (SCT) at cloud base 2,000 feet, temperature 27°C with due point 25°C and aerodrome pressure was 1010 mbs.

4. Weight and Balance

a. Total fuel 6,500 kg
b. Actual take-off weight 63,974 kg
c. CG18.57% MAC
d. Landing weight 62,000 kg

The first scenario was executed to observe a normal takeoff (no malfunction) followed by an approach to stall at flaps up (flaps 0) with mid-range thrust, flaps up with idle thrust and flaps 5 with higher thrust (similar to a departure stall).

The first scenario consisted of four cases to observe normal MCAS activation. General observation revealed that MCAS function activation in the flaps up configuration was similar in fashion to Speed Trim System in the flaps down configuration and that it operated at the same rate.

The second scenario was executed in the same simulator setup condition with addition of speed (SPD) and altitude (ALT) failure flag simulation. The observation noted that no fault or caution lights were illuminated, however, if the recall button was pressed, the SPEED TRIM FAIL and MACH TRIM FAIL lights would illuminate.

The third scenario was executed with the objective to observe AOA DISAGREE. The scenario was conducted in four cases.

The first case objective of the third scenario was to observe an "ideal" response to stick shaker and IAS and ALT DISAGREE messages after takeoff. Crews were instructed to maintain takeoff flap configuration (MCAS function will not engage with flaps extended), complete the Airspeed Unreliable and Altitude Disagree NNC and return for landing at Jakarta. There was no significant issue regarding the first case.

The second case of the third scenario was to observe the activation of MCAS, completion of Airspeed Unreliable, Altitude Disagree and Runway Stabilizer checklists, and the effects of performing the Runaway Stabilizer checklist on MCAS, similar to Lion Air flight 043. Crews were instructed to clean up the aircraft after takeoff, complete the Airspeed Unreliable, Altitude Disagree and Runaway Stabilizer checklists and return for landing at Jakarta using manual trim. The observation noticed that the MCAS activation in the flap up configuration made the flying more difficult. The aircraft was still controllable, as long as the flight crew countered the MCAS activation using the control column trim switches. The problem was solved using the runaway stabilizer procedure.

During the execution of second case of the third scenario, the AOA bias was set on the right AOA sensor prior the flight and the FCC A remained as master. As a result, no erroneous MCAS activations were expected, however, the left stick shaker activated and ALT DISAGREE and IAS DISAGREE was noticed. The observation found that the erroneous airspeed was quickly notice by identifying the pitch limit indicator position on the affected PFD.

The third case of the third scenario was to observe the conditions that occurred during the accident flight. The flight crews were instructed to clean up the aircraft after takeoff, begin the Airspeed Unreliable and Altitude Disagree NNCs, and respond to ATC requests. The control column electric trim was initially used to counter MCAS, then insufficient control column electric trim was used similar to the accident flight. The observations were as follows:

1. Altitude was not able to be maintained with aft control column force if short activations of electric trim result in an accumulation of mis-trim from MCAS nose down trim commands.

2. Repeated MCAS activations increased the flight crew workload and required more attention to counter it.

3. The recreated accident flight ATC communication was distracting.

4. It was hard to get through the Airspeed Unreliable NNC. Being unable to complete the airspeed unreliable NNC and/or not being able to identify reliable airspeed/altitude increased the flight crew work load..

5. Discussion arose regarding flight crew request to ATC to confirm altitude. This behavior indicated that the flight crew and the ATC likely did not realize that the altitude displayed on the radar was almost certainly Mode C / Mode S data from the aircraft and was no more reliable than what was displayed in the flight deck. Hearing ATC providing an altitude might cause the flight crew to consider the altimeter was reliable when trying to diagnose the problem. A comment was made that prior accidents had involved flight crews requesting altitude information from ATC. Suggestions made included additional information/training to flight crews and possibly have the controller reply to such requests with, "Your aircraft is reporting altitude as …. feet" to remind flight crew how the system works. .

The fourth case of the third scenario was for the flight crew to feel the control column forces during single and repetitive MCAS activations while trying to keep the aircraft level. The AOA bias was introduced and the MCAS function activated. The observations were as follows:

1. Significant aft control column force was necessary to hold the control column after one activation of MCAS.

2. After two full applications of MCAS and no restoring electric manual trim up, one participant characterized the control column force as "too heavy."

CHAPTER 4

ORGANIZATIONAL AND MANAGEMENT INFORMATION

Aircraft Operator

The PK-LQP aircraft is operated by PT. Lion Mentari Airlines (Lion Air) which had valid Air Operator Certificate (AOC) number 121-010.

The Lion Air operates a total of 120 aircraft consisting of 3 Airbus A330, 1 Boeing 747-400, 1 Boeing 737-400, 37 Boeing 737-800, 67 Boeing 737-900ER and 11 Boeing 737-8 (MAX) (including the accident aircraft) and serves more than 120 destinations with up to 630 flights daily.

The Lion Air has several manuals contains policy and procedure approved by the Directorate General of Civil Aviation. The relevant policy and procedure are described in the following subchapter.

Operation Manual (OM)-part A

Captain Duty and Responsibility
Some of the Captain responsibilities according to the Operation Manual (OM)-part A subchapter 1.4, is directly and specifically responsible for, and is the final authority as to, the operation of the aircraft. Therefore, the Captain is responsible for ensuring the aircraft is in condition for safe flight and must discontinue the flight when un-airworthy mechanical, electrical, or structural conditions occur.

In regards with defect report, the OM-part A subchapter 2.1.14.3 described that Captain has following responsibility:

• Before the flight, Captain must examine the Aircraft Flight and Maintenance Log (AFML) to inquire about the technical status of the aircraft.

• Ensure that AFML is filled out completely and correctly.

•Record and report all defects in AFML and ensure all information entered in AFML:

- is up to date;

- is legible (write clearly);

- cannot be erased (using ball point); and

- is correctable in the case of an error provided each correction is identifiable and errors remain legible (cross the error, write correction and put Captain sign next to it).

•If the content in the AFML is considered void, Captain must cross the page and put word VOID along with the crossed page.

•After the flight, Captain must ensure that all discrepancies and mechanical irregularities noted during the course of a flight or series of flights are entered in the AFML. Where applicable, snags entered in the Cabin Maintenance log, which are airworthiness items, must be transferred by the Flight Crew into the AFML. In addition, and where possible, Captain must debrief maintenance personnel directly regarding reported aircraft defects.

The OM-part A did not describe flight crew requirement to report faults to include fault code which refer to Fault Reporting Manual (FRM) nor did it describe any requirement for a crewmember to report using a guarded switch in flight.

Duty Management Pilot, Duty and Responsibility

The OM-part A subchapter 1.3.2.2 described Duty Management Pilot is the management flight crew rostered outside normal office hours representing Flight Operation Department, who is responsible directly to the Director of Flight Operation. The roster for work day is from 1730 to 0830 LT the following morning, while for weekend (Saturday and Sunday) and public holiday is comprising two different duty periods, which are from 0830 to 1730 LT and continue from 1730 to 0830 LT the following morning.

The Duty Management Pilot has to be located at designated residential address or other reasonable notified address with contactable and uninterrupted telephone, company group network (online messaging group application) and/or email communication. The Duty Management Pilot has duties and responsibilities to handle and effective management of flight/schedule disruptions and delays for resolution in a timely manner by coordination with MCC, Station Manager and other respective departmental focal points. The Duty Management Pilot also has responsibility as Liaison and coordination with Chief Pilot, Fleet Managers, Head of Training, Safety and Security Manager, Flight Data Analysis Manager, etc., as required for urgent safety attention, critical high severity or airworthiness matters of concern.

Handover Control Procedures

OM-part A described procedure related to the handover control as follows:

During handover and undertaking of flight control, the following phraseology must be used in order to make the transfer clear:

• *Pilot handing over the control:* "YOU HAVE CONTROL"

• Pilot undertaking the control: "I HAVE CONTROL"

Flight crew who handed over the control must convert to PM after normal control of airplane, by the flight crew who has undertaken the control is confirmed.

Any change over of control from Second in Command to the PIC should be done any time the PIC deems it necessary.

Urgency and Distress Condition Procedures

The OM-part A "Communication with ATC" describes distress condition as a condition of being threatened by serious and/or imminent danger and requires immediate assistance. The distress call must use radiotelephony signal of MAYDAY at the commencement of the first distress communication and has absolute priority over all other radio traffic, other stations shall not transmit on the frequency concerned until the distress communication is ended or transferred to another frequency.

While the urgency signal uses PAN PAN. The urgency condition was described as a condition concerning the safety of an aircraft or other vehicle, or of some person on board or within sight, but which does not require immediate assistance.

Urgency communication has priority over has priority over all other communications, except distress communication. Other station shall not interfere with the transmission of urgency traffic.

The procedures related to the urgency and distress condition described in the OM-part A subchapter was in accordance with ICAO Annex 10 subchapter 5.3.

The duty assignment for Pilot Flying (PF) and Pilot Monitoring (PM) in abnormal and emergency situations were described in the following assignment of tasks sharing is recommended:

EMERGENCY AND ABNORMAL SITUATIONS

PF IS RESPONSIBLE FOR	PM IS RESPONSIBLE FOR:
• Throttles; • Flight path and airspeed control; • Aircraft configuration; • Navigation; • Communication.	• Checklist reading; • *Execution of required actions on PF request;* • *Engine fuel levers, fire handles and guarded switches (irreversible actions), with confirmation of PF.*

Emergency and abnormal procedures are to be initiated on command of the Pilot Flying.

The Pilot in Command may change over the control at any time to ensure that the highest level of safety is maintained.

During an emergency or abnormal situation, the Pilot in Command must allocate crew duties to ensure that the highest level of situation awareness is maintained in the cockpit and cabin. This will prevent all attention being totally directed at resolving the emergency or abnormal situation to the detriment of safe flight. Any ambiguities, confusion, unresolved discrepancies or use of improper procedures must be discussed immediately, and if necessary, a missed approach initiated to allow remedial action at safe altitude.

Following a flaps malfunction / jamming, any approach and landing with zero-degree flap configuration must be flown by the Pilot in Command. Approach and landing following other emergency or abnormal situation must be conducted such as to ensure the highest level of safety.

Following an in-flight emergency or abnormal situation, all approach either instrument or visual should not be commenced or should be discontinued, until the Emergency Memory Items and subsequent procedures and have been completed. For more detail, refer to respective aircraft type FCOM.

The duty assignment for PF and PM in normal situation described in OM-part A subchapter. The following assignment of tasks sharing is recommended:

NORMAL SITUATIONS

PF IS RESPONSIBLE FOR	PM IS RESPONSIBLE FOR:
• Flight path and airspeed control; • Aircraft configuration; • Navigation.	• Flight path and airspeed control monitoring; • Navigation monitoring; • Communication; • Checklist reading; • Execution of actions on PF request.

Hazard and Occurrence Report Procedures

The OM-part A subchapter required Pilot in Command to report the occurrence that have or may have safety impact on operation procedure by filling an Air Operations - Safety Hazards & Occurrence Report (A-SHOR) and for all Lion Air personnel are also encouraged to voluntarily report any error or hazards that may have led to incident or accident by filling a Safety Hazard and Occurrence Report (SHOR).

The subchapter described serious incident as an incident involving circumstances indicating that there was a high probability of an accident and associated with the operation of an aircraft. This subchapter provided several examples of serious incident, including:

• *Malfunction or failure of the flight control system; and*

• *System failures, weather phenomena, operations outside the approved flight envelope or other occurrences which caused or could have caused difficulties controlling the aircraft*

The subchapter described:

In the event of an accident or a serious incident, either airborne or on the ground, the Pilot in Command or a crew member, if physically able, or any other person will advise Operational Control Department (OCD) by the quickest available means, that will in turn advise Safety & Security Directorate.

In the case the OCD is aware of a LION AIR aircraft accident or a serious incident or, has reasons to believe a LION AIR aircraft has been involved in an accident, or in the case of an overdue aircraft report, the OCD will immediately advise LION AIR Safety & Security Directorate by the quickest available means.

The subchapter described that Pilot in Command has responsibilities to notify the SS Directorate when the aircraft is believed have been experienced accident or serious incident. Afterwards, the SS Directorate must immediately, and by the most suitable and quickest means available, notifying several related authorities including the *Komite Nasional Keselamatan Transportasi.*

The subchapter described all operational personnel also encourage reporting to company of non-exhaustive list of events, which included in part:
- *Aircraft handling difficulties including abrupt maneuver, excessive pitch attitude, aircraft trim problems, un-commanded roll, or un-commanded turn;*

- *Warning or alert, including flight control warnings, door warnings, stall warning (stick-shaker), fire/smoke/fumes warning and stall or stall warning.*

The OM-part A subchapter described methods to report hazard and occurrence to SS Directorate as follows:
a. Primary reporting method is using web-based application software which provides online form of Air Operations - Safety Hazards & Occurrence Report (A-SHOR).

b. Secondary reporting method is to fill a paper form of A-SHOR and send the form to the Safety and Security (SS) Directorate via email.

c. If the primary and secondary methods are not practicable, the reporter can use facsimile to send the A-SHOR or short message service to SS Directorate.

Minimum Equipment List Procedures
OM-part A subchapter 8.6.9 described the applicability of Minimum Equipment List (MEL) as follows:

The provisions of the MEL are applicable until the aircraft commences the flight (i.e. when the aircraft begins to move under its own power for take-off).

Failures occurring between "Off Blocks"[13] and "Brake Release" require accomplishment of the appropriate abnormal procedure by the crew. Upon completion of the procedure, once the failure has been clearly identified and confirmed, and prior to take-off, the MEL must however be consulted:

- *If the item is NO GO[14] or if the MEL requires the completion of a maintenance procedure, the PIC must return to the blocks;*

- *If the item is "GO" or "GO IF" in the MEL, any decision to continue the flight must be subject to Pilot in Command judgment and good airmanship taking into account all other relevant factors, performance penalties and operational restrictions related to the intended flight.*

The MEL may also be consulted in flight to decide on an in-flight turn back depending on the "GO"/ "NO GO" status of the failed function and the possibility to repair the failure at the destination

Operation Training Manual
The Operation Training Manual subchapter stated:

UNSATISFACTORY PERFORMANCE

In the event a pilot fails to demonstrate competence in a section of a check or does not satisfactorily complete a check, Check Airman (CCP) notifications must be made immediately to the Crew Scheduling Officer, Head of Training, Chief Pilot and DO to ensure that the crew member is not assigned to operate, nor does he operate, any Company aircraft or flight, until he has satisfactorily been re-tested for the section or whole of the check, as applicable.

When a CCP decides that a pilot has failed during course of a check, the check shall be terminated. The time remaining in the session may be used as training provided that the candidate is advised at the time of failure upon the completion of the training flight the candidate is debriefed on the reason of the failure.

Following an unsatisfactory Base Training/ Check, the pilot shall not be submitted to a series of re-test in the item(s) concerned until he has undergone a period of training and has attained a satisfactory standard during that training.

For Administration Refer to Supplemental/ Additional Training (Simulator or Flying).

SUPPLEMENTAL/ ADDITIONAL TRAINING (SIMULATOR OR FLYING)

GENERAL

In addition to the training described elsewhere in this manual, Supplemental/Additional Training will be carried out in the following cases:

- *Supplemental Training after a failure of a Base Check, Periodic Proficiency Check, Line Check or Annual line Check;*

- *Additional Training during Conversion Training or after unsatisfactory performance during Recurrent Training/LOFT.*

- *The pilot has been through an accident/incident and required by the DGCA to perform corrective training.*

- Recommendation from fleet and SSQ as a result of FDA accident.

The fleet determines the type and amount of training to be administered to the Pilot.

Supplemental/ Additional Training are subject to approval by Head of Training, however he/she may delegate the authority to respective Fleet Training Manager. In certain situations, the Training Review Board may convene to decide whether supplemental/ additional training should be approved and, if approved, the amount of training.

Based on the discussion with the Lion Air management, the Lion Air policy for flight crew assessed of having inability to perform minimum requirement during training was that the flight crew would be treated with briefing or rehearsal.

The training manual described the policy for flight crew assess as unsatisfactory or fail the test. However, the investigation could not find the policy for flight crew assessed of having inability to perform minimum requirement for single assessment item which general result is satisfactory.

Flight Crew Operations Manual (FCOM)
Normal Procedure Chapter, Section 21 "Amplified Procedures" section "Preliminary Pre-flight Captain or First Officer" included in part: …

Maintenance documents……………………Check

MAINT light……………….Verify extinguished

STALL WARNING TEST switchers. Push and hold, one at a time
Verify that each control column vibrates when the respective switch is pushed.

Note: *The stall warning test requires that AC transfer busses are powered for up to 4 minutes.*

Note: *With hydraulic power off, the leading edge flaps can droop enough to cause an asymmetry signal, resulting in a failure of the stall warning system test. Should this* occur, obtain a clearance to pressurize the hydraulic system, place the "B" system electric pump ON and retract the flaps. When flaps are retracted repeat the test. At the completion of the test, turn the "B" system electric pump "OFF".

Also in that same chapter, the section titled "Preflight Procedure – Captain" stated in part:

> *Engine start levers...............................CUTOFF*
>
> *STABILIZER TRIM cutout switches...........NORMAL*

Appendix, Section 1.13.1 "Overall Philosophy – Policy", stated in part *"...Standard calls enhance crew awareness and co-ordination, and are standardized to remove ambiguity... They must be used in all normal and non-normal operation...The broad policy regarding standard calls in flight is that the Pilot Flying (PF) shall make all FMA calls and requests for mode changes, configuration changes, checklists and decision calls, as he is responsible for the aircraft flight path. The Pilot Monitoring (PM) shall make all other calls since they are to assist with the PF's situational awareness. If a call is missed by the responsible crewmember, it shall be made by the other."*

Section 1.13.3 "Standard Calls" provided the following guidance:

Pilot Flying
The following request calls are made by the Pilot Flying (PF):

1. Configuration requests
2. Thrust requests
3. MCP requests
4. Checklist requests

Decision calls are made by the PF. They are in effect statements of intention. The compulsory calls are listed below. It must be appreciated that any other decision calls can be made by the PF if he thinks that the information is important and enhances CRM. This becomes particularly relevant during non- normal procedures.

Pilot Monitoring
The Pilot Monitoring (PM) will make all the following calls:

1. Altitude calls
2. Speed calls
3. Approach Parameter Deviation calls
4. Instrument calls

5. Lighting calls

If a call is valid and understood, it is acknowledged by the corresponding crewmember with the appropriate response. This does not apply to the V1 and VR calls as the response is a standard action. If a standard call or FMA call is responded to by another standard call or FMA call, the response "CHECK" is omitted. Calls made by the auto-callout system are not to be made by the PM unless the system is inoperative or fails to make the call.

Flight Crew Training Manual (FCTM)

The Flight Crew Training Manual, Chapter 3 "Takeoff and Initial Climb" stated in part, the following:

Rotation and Liftoff - All Engines

Takeoff speeds are established based on minimum control speed, stall speed, and tail clearance margins. Shorter-bodied airplanes are normally governed by stall speed margin while longer-bodied airplanes are normally limited by tail clearance margin. When a smooth continuous rotation is initiated at VR, tail clearance margin is assured because computed takeoff speeds depicted in the PI Chapter of the FCOM, airport analysis, or FMC, are developed to provide adequate tail clearance.

Above 80 knots, relax the forward control column pressure to the neutral position. For optimum takeoff and initial climb performance, initiate a smooth continuous rotation at VR toward 15° of pitch attitude. However, takeoffs at low thrust setting (low excess energy) will result in a lower initial pitch attitude target to achieve the desired climb speed.

The use of stabilizer trim during rotation is not recommended. After liftoff, use the attitude indicator, or indications on the PFD or HUD (HUD equipped airplanes), as the primary pitch reference. The flight director, in conjunction with indicated airspeed and other flight instruments is used to maintain the proper vertical flight path.

Note: *The flight director pitch command is not used for rotation.*

With a consistent rotation technique, where the pilot uses approximately equal control forces and similar visual cues, the resultant rotation rate differs slightly depending upon airplane body length.

Note: *Do not adjust takeoff speeds or control forces to compensate for increased body length.*

Using the technique above, resultant rotation rates vary from 2° to 3° per second with rates being lowest on longer airplanes. Liftoff attitude is achieved in approximately 3 to 4 seconds depending on airplane weight and thrust setting.

Chapter 8 "Non normal operations" stated in part:

Aircrews are expected to accomplish Non-Normal Checklists (NNCs) listed in the QRH. These checklists ensure maximum safety until appropriate actions are completed and a safe landing is accomplished. Techniques discussed in this chapter minimize workload, improve crew coordination, enhance safety, and provide a basis for standardization.

When a non-normal situation occurs, the following guidelines apply:

☐ *Non-normal recognition: The crewmember recognizing the malfunction calls it out clearly and precisely*

☐ *Maintain airplane control: It is mandatory that the Pilot Flying (PF) fly the airplane while the Pilot Monitoring (PM) accomplishes the NNC. Maximum use of the auto-flight system is recommended to reduce crew workload*

☐ *Analyze the situation: NNCs should be accomplished only after the malfunctioning system has been positively identified. Review all caution and warning lights to positively identify the malfunctioning system(s)*

☐ *Take the proper action: Although some in-flight non-normal situations require immediate corrective action, difficulties can be compounded by the rate the PF issues commands and the speed of execution by the PM. Commands must be clear and concise, allowing time for acknowledgment of each command prior to issuing further commands. The PF must exercise positive control by allowing time for acknowledgment and execution. The other crewmembers must be certain their reports to the PF are clear and concise, neither exaggerating nor understating the nature of the non-normal situation. This eliminates confusion and ensures efficient, effective, and expeditious handling of the non-normal situation*

☐ *Evaluate the need to land: If the NNC directs the crew to plan to land at the nearest suitable airport, or if the situation is so identified in the QRH, diversion to the nearest airport where a safe landing can be accomplished is required. If the NNC or the Checklist Instructions do not direct landing at the nearest suitable airport, the pilot must determine if continued flight to destination may compromise safety.*

Landing at the Nearest Suitable Airport

In a non-normal situation pilot-in-command, having the authority and responsibility for operation and safety of the flight, must make the decision to continue the flight as planned or divert. In an emergency situation, this authority may include necessary deviations from any regulation to meet the emergency. In all cases, the pilot-in-command is expected to take a safe course of action.

The QRH assists flight crews in the decision making process by indicationg thosesituations where "landing at the nearest suitable airport" is required... If required to divert to the nearest suitable airport, the guidance material typically specifies that the pilot should select the nearest suitable airport "in point of time" or "in terms of time." In selecting the nearest suitable airport, the pilot-in-command should consider the suitability of nearby airports in terms of facilities and weather and their proximity to the airplane position. The pilot-in-command may determine, based on the nature of the situation and an examination of the relevant factors, that the safest course of action is to divert to a more distant airport than the nearest airport...

Flight Instruments, Displays: Airspeed Unreliable

...Increased reliance on automation has de-emphasized the practice of setting known pitch attitudes and thrust settings. However, should an airspeed unreliable incident occur, the flight crew should be familiar with the approximate pitch attitude and thrust setting for each phase of flight. This familiarity can be gained by noting the pitch attitude and thrust setting occasionally during normal flight. Any significant change in attitude from the attitude normally required to maintain a particular airspeed or Mach number should alert the flight crew to a potential airspeed problem.

If abnormal airspeed is recognized, immediately set the target pitch attitude and thrust setting for the aircraft configuration from the Airspeed Unreliable memory items. When airplane control is established, accomplish the Airspeed Unreliable NNC. The crew should alert ATC if unable to maintain assigned altitude or if altitude indications are unreliable. The following assumptions and requirements were used in developing these memory items:

- The memorized settings are calculated to work for all model/engine combinations, at all weights and at all altitudes.
- The flaps up settings will be sufficient such that the actual airspeed remains above stick shaker and below overspeed.
- The flaps extended settings will be sufficient such that the actual airspeed remains above stick shaker and below the flap placard limit.
- The settings are biased toward a higher airspeed as it is better to be at a high energy state than a low energy state.
- These memorized setting are to allow time to stabilize the airplane, remain within the flight envelope without overspeed or stall, and then continue with reference to the checklist.

- *Settings are provided for flight with and without flaps extended. The crew should use the settings for the condition they are in to keep the airplane safe while accessing the checklist.*

The memorized pitch and thrust setting for the current configuration (flaps extended/flaps up) should be applied immediately with the following considerations:

- *The flaps extended pitch and thrust settings will result in a climb.*

- *The flaps up pitch and thrust settings will result in a slight climb at light weights and low altitudes, and a slight descent at heavy weights and high altitudes.*

- *At light weight and low altitude, the true airspeed will be higher than normal, but within the flight envelope. At heavy weight and high altitude, the same settings will result in airspeed lower than normal cruise but within the flight envelope.*

- *The goal of these pitch and power settings is to maintain the airplane safely within the flight envelope, not to maintain a specific climb or level flight.*

- *The current flap position should be maintained until the memory pitch and thrust settings have been set and the airplane stabilized. If further flap extension/flap retraction is required refer to PI-QRH Airspeed Unreliable table.*

In order to determine if a reliable source of indicated airspeed is available, the Airspeed Unreliable checklist says "When in trim and stabilized, cross check the captain, first officer and standby airspeed indicators." The intent of this statement is for the pilot flying to set the pitch attitude and thrust setting from the PI-QRH Flight With Unreliable Airspeed table and allow the airplane to stabilize before comparing the airspeed indications to those shown in the table.

The airplane is considered stabilized when the thrust and pitch have been set, and the pitch is trimmed with no further trim movement needed to maintain the pitch setting. This is not an instantaneous process, and must be complete before comparing indicated and expected airspeeds for accurate results.

If it is determined that none of the airspeed indicators are reliable, the PI-QRH tables should be used for the remainder of the flight. Flight crews need to ensure they are using the table and values appropriate for phase of flight and airplane configuration.

- When changing phase of flight or airplane configuration, make initial thrust change, set pitch attitude, configure the airplane as needed, then recheck thrust and pitch, and trim as needed. Do not change configuration until the airplane is trimmed and stabilized at the current configuration.

If the flight crew is aware of the problem, flight without the benefit of valid airspeed information can be safely conducted and should present little difficulty. Early recognition of erroneous airspeed indications requires familiarity with the interrelationship of attitude, thrust setting, and airspeed. A delay in recognition could result in loss of airplane control.

Ground speed information is available from the FMC and on the instrument displays. These indications can be used as a crosscheck. Many air traffic control radars can also measure ground speed.

For airplanes equipped with an Angle of Attack (AOA) indicator, maintain the analog needle at approximately the three o'clock position. This approximates a safe maneuver speed or approach speed for the existing airplane configuratio

Quick Reference Handbook (QRH)

The Lion Air 737 Flight Crew Operations Manual. Checklist instructions. Chapter CI. Section , dated June 15, 2017 stated in part, the following:

The non-normal checklists chapter contains checklists used by the flight crew to manage non-normal situations...

Most checklists correspond to a light, alert or other indication. In most cases, the MASTER CAUTION and system annunciator lights also illuminate to indicate the non-normal condition.

All checklists have condition statements. The condition statement briefly describes the situation that caused the light, alert or other indication. Un-annunciated checklists also have condition statements to help in understanding the reason for the checklist.
Some checklists have objective statements. The objective statement briefly describes the expected result of doing the checklist or briefly describes the reason for steps in the check list.

Check lists can have both memory and reference items. Memory items are critical steps that must be done before reading the checklist. The last memory item is followed by a dashed horizontal line. Reference items are actions to be done while reading the checklist.

Some checklists have additional information at the end of the checklist. The additional information provides data the crew may wish to consider. The additional information does not need to be read.

Checklists that need a quick response are listed in the Quick Action Index which also available on the QRH cover page. In each system section, Quick Action Index checklists are listed first, followed by checklists that are not in the Quick Action Index. The titles of Quick Action Index checklists are printed in bold type. Checklist titles in upper case (such as AUTO BRAKE DISARM) are annunciated by a light, alert, or other indication. Checklist titles in upper and lower case (such as Window Damage) are not annunciated.

Non-Normal Checklist Operation

Non-normal checklists start with steps to correct the situation. If needed, information for planning the rest of the flight is included…

While every attempt is made to supply needed non–normal checklists, it is not possible to develop checklists for all conceivable situations. In some multiple failure situations, the flight crew may need to combine the elements of more than one checklist. In all situations, the captain must assess the situation and use good judgment to determine the safest course of action.

It should be noted that, in determining the safest course of action, troubleshooting, i.e., taking steps beyond published non-normal checklist steps, may cause further loss of system function or system failure. Troubleshooting should only be considered when completion of the published non-normal checklist results in an unacceptable situation.
There are some situations where the flight crew must land at the nearest suitable airport. These situations include, but are not limited to, conditions where:

- *the non–normal checklist includes the item "Plan to land at the nearest suitable airport."*
- *fire or smoke continues*
- *only one AC power source remains (engine or APU generator) only one hydraulic system remains (the standby system is considered a hydraulic system)*
- *any other situation determined by the flight crew to have a significant adverse effect on safety if the flight is continued.*

Non–normal checklists also assume:

- *During engine start and before takeoff, the associated non–normal checklist is done if a non-normal situation is identified. After completion of the checklist, the Dispatch Deviations Guide or operator equivalent is consulted to determine if Minimum Equipment List dispatch relief is available.*

- *System controls are in the normal configuration for the phase of flight before the start of the non-normal checklist.*

- *If the MASTER CAUTION and system annunciator lights illuminate, all related amber lights are reviewed to assist in recognizing the cause(s) of the alert.*

- *Aural alerts are silenced and the master caution system is reset by the flight crew as soon as the cause of the alert is recognized.*

- *The EMERGENCY position of the oxygen regulator is used when needed to supply positive pressure in the masks and goggles to remove contaminants. The 100% position of the oxygen regulator is used when positive pressure is not needed but contamination of the flight deck air exists. The Normal position of the oxygen regulator is used if prolonged use is needed and the situation allows. Normal boom microphone operation is restored when oxygen is no longer in use.*

- *Indicator lights are tested to verify suspected faults.*

- *In flight, reset of a tripped circuit breaker is not recommended. However, a tripped circuit breaker may be reset once, after a short cooling period (approximately 2 minutes), if in the judgment of the captain, the situation resulting from the circuit breaker trip has a significant adverse effect on safety. On the ground, flight crew reset of a tripped circuit breaker should only be done after maintenance has determined that it is safe to reset the circuit breaker.*

- *Flight crew cycling (pulling and resetting) of a circuit breaker to clear a non-normal condition is not recommended, unless directed by a non-normal checklist.*

Non-Normal Checklist Use

Non-normal checklist use starts when the airplane flight path and configuration are correctly established. Only a few situations need an immediate response (such as CABIN ALTITUDE WARNING or Rapid Depressurization). Usually, time is available to assess the situation before corrective action is started. All actions must then be coordinated under the captain's supervision and done in a deliberate, systematic manner. Flight path control must never be compromised.

When a non-normal situation occurs, at the direction of the pilot flying, both crewmembers do all memory items in their areas of responsibility without delay. The pilot flying calls for the checklist when:

- the flight path is under control

  - *the airplane is not in a critical phase of flight (such as takeoff or landing)*
  - *all memory items are complete.*

The pilot monitoring reads aloud:

- the checklist title
- *the airplane effectivity (if applicable) as needed to verify the correct checklist*
- *as much of the condition statement as needed to verify that the correct checklist has been selected*
- *as much of the objective statement (if applicable) as needed to understand the expected result of doing the checklist.*

The pilot flying does not need to repeat this information but must acknowledge that the information was heard and understood.

For checklists with memory items, the pilot monitoring first verifies that each memory item has been done. The checklist is normally read aloud during this verification. The pilot flying does not need to respond except for items that are not in agreement with the checklist. The item numbers do not need to be read....

Non-memory items are called reference items. The pilot monitoring reads aloud the reference items, including:

- - the precaution (if any)
- - the response or action
- - *any amplifying information.*

The pilot flying does not need to repeat this information but must acknowledge that the information was heard and understood.

For checklists with memory items, the pilot monitoring first verifies that each memory item has been done. The checklist is normally read aloud during this verification. The pilot flying does not need to respond except for items that are not in agreement with the checklist. The item numbers do not need to be read....

Non-memory items are called reference items. The pilot monitoring reads aloud the reference items, including:

- the precaution (if any)
- the response or action
- *any amplifying information.*

The pilot flying does not need to repeat this information but must acknowledge that the information was heard and understood. The item numbers do not need to be read.

With the airplane in flight or in motion on the ground the pilot flying and the pilot monitoring take action based on each crewmember's Areas of Responsibility. After moving the control, the crewmember taking the action also states the checklist response.

The pilot flying may also direct reference checklists to be done by memory if no hazard is created by such action, or if the situation does not allow reference to the checklist.

Each checklist has a checklist complete symbol at the end. The following symbol indicates that the checklist is complete:

■ ■ ■ ■

After completion of each non–normal checklist, the pilot monitoring states "___ CHECKLIST COMPLETE."

Fault Reporting Manual (FRM)

The Boeing Company provides Fault Reporting Manual (FRM) and the Interactive Fault Isolation Manual (IFIM) which together provides a structured method for the aircraft operator to report and correct faults in aircraft systems. The manuals are linked by a unique 8-digit fault code for each fault.

The FRM is primarily for the flight crews and the cabin crews and contains three alphabetical lists of faults, with a fault code for each fault. While the IFIM is for the maintenance crews which gives a fault isolation procedure for each of the faults.

The faults divided as:

A. Observed Faults are the problem symptoms (other than Status messages) that can be observed by the flight crew or the maintenance crew. These include:

(a) Fault alerts and annunciations on the pilot's display
(b) Engine exceedances on the pilot's display
(c) Failure flags on the navigation display
(d) Flight crew observations in the flight deck
(e) Flight crew observations during the pre-flight inspections
(f) Servicing crew observations
(g) Cargo loading crew observations
(h) Ground maintenance crew observations.

B. Status Messages are the status messages show on the pilots' status display.

C. Cabin Faults are the problem symptoms which can occur with the systems and equipment in the passenger cabin.

The FRM can be used to report faults by the flight crew when a fault occurs. The fault code can be found on the applicable section and look for the text which matches the fault. The flight crew or cabin crew shall write the fault code in the aircraft log book.

The fault code can be sent ahead to the destination before the aircraft lands, to allow the maintenance crew to prepare for isolation of the fault by voice radio or Aircraft Communications Addressing and Reporting System (ACARS).

Minimum Equipment List

At the time of the occurrence, the Lion Air MEL for Boeing 737-8 (MAX) identified as document number LA-CM-04-07 published on revision 02 issued on 16 April 2018 which was approved by Indonesia DGCA on 22 May 2018. The Lion Air MEL of Boeing 737-8 MAX was referring to the Boeing Master Minimum Equipment List (MMEL) revision 1.

In the preamble of the Lion Air MEL, the equipment related to the airworthiness and the operating regulations of the aircraft is required to be operative. The statement is as follows:

A Master Minimum Equipment List (MMEL) is developed by the FAA, with participation by the aviation industry, to improve aircraft utilization and thereby provide more convenient and economic air transport for the public.

The FAA approved MMEL includes those items of equipment related to airworthiness and operating regulations and other items of equipment which Administrator finds may be inoperative and yet maintain an acceptable level of safety by appropriate conditions and limitations; it does not contain obviously required items such as wings, flaps, and rudders.

It is important to remember that all equipment related to the airworthiness and the operating regulations of the aircraft not listed on the PT. LION AIR MEL must be operative.

The Boeing 737-8 (MAX) MEL did not list the indicated airspeed and altimeter.

Safety Management System Manual

The Lion Air safety policy dated 15 January 2015, in the Safety Management System Manual (SMSM), dated 01 February 2018, regarding safety report described that senior management is committed to:

• *Mandating and facilitating safety as major part of individual responsibility involving all departments, business partners, contractors, and suppliers with no exception, whatever their position and hierarchical status.*

• *Performing the process safety risk and hazard management associated with the company operations through the implementation and continuous improvement of Safety Management System.*

• *Evaluating the achievement level periodically using measured performance indicators against the pre-determined realistic objectives and/or targets through management reviews.*

• *Ensuring that all staff is provided with adequate and appropriate information, training, competency, and provision of necessary resources of safety standards.*

• *Developing and encouraging the safety reporting culture to all employees by 'Non-Punitive' policy that will guarantee no adverse action taken against them. However, illegal activity, willful or deliberate violations will not be tolerated.*

This Safety Policy reflects the company commitment to the safety culture and issued under the authority of the highest level of management in the organization

The SMSM subchapter 1.3 defines hazard as:

Means condition, object or activity with the potential of causing injuries to personnel, damage to equipment or structures, loss of material, or reduction of ability to perform a prescribed function. Mitigation means measures to address the potential hazard or to reduce the risk probability or severity.

That same section defines just culture as:

Means a culture in which front-line operators or other persons are not punished for actions, omissions or decisions taken by them that are commensurate with their experience and training, but in which gross negligence, willful violations and destructive acts are not tolerated.

The SMSM subchapter 3.1 described that:

Lion Air has operational safety reporting system that is implemented throughout the organization in a manner that:

• *Encourages and facilitates personnel to submit reports that identify safety hazards, expose safety deficiencies and raise safety concerns;*

• *Ensures mandatory reporting in accordance with applicable regulations;*

• *Includes analysis and management action as necessary to address safety issues identified through the reporting system.*

Lion Air Safety management systems involve the reactive and proactive identification of safety hazards. Reporting systems is not just restricted to incidents but include provision for the reporting of hazards, i.e. unsafe conditions that have not yet caused an incident. Lion Air maintains a just reporting policy and encourages the utilization of the reporting system for the purpose for which it is intended. To identify and reduce the hazards associated within the aviation industry. It is everyone's responsibility to report hazards, occurrences, or incidents that may become an accident.

Reporting system will facilitate and encourages the reporting of hazards, deficiencies and safety concerns from personnel at all levels of the organization. Lion Air recognize that an acknowledgement for each report is essential to build and maintain confidence in the process and encourage more reporting from all personnel within the company.

All personnel may report any hazard that has the potential to cause damage or injury or that threatens the organizations viability. Hazards and incidents should be reported if it is believed that something can be done to improve safety, other aviation personnel could learn from the report, the system and its inherent defences did not work "as advertised".

SMSM subchapter 3.2 stated in part, ...*all employees are encouraged to report anything that they perceive as a hazard or a threat to safety by any means of communication. It is the Safety and Security (SS) Department obligation to validate the report and provides a mechanism for a review and analysis of each report to determine whether a real safety issue exists, and if so, ensure development and implementation of appropriate action by responsible operational managers to correct the situation.*

In the SMSM subchapter 3.3 described occurrences must be reported to the SS Department reporting system is as follows:

a. General occurrence, which included accident, incident and hazard.

    In this subchapter, hazard defined as: *a situation or condition which, if unchecked, could lead to a negative outcome. A hazard is not the negative outcome itself, i.e. it is not an incident or accident.*

b. Flight operations and dispatch safety occurrence, which included examples of:

    • Aircraft handling difficulties including abrupt maneuver, excessive pitch attitude, aircraft trim problems, un-commanded roll, or un-commanded turn;

    • Warning System and Air Navigation Service event including warning or alert, including flight control warnings, door warnings, stall warning (stick-shaker), fire/smoke/fumes warning and stall or stall warning.

c. Cabin operation safety occurrence;

d. Ground operations and cargo safety occurrence;

e. Maintenance and engineering safety events.

The SMSM subchapter 3.4 also described the method to report hazard and occurrence to SS Directorate as follows:

a. Web-based application software which allowed reporter to fill online form of Safety Hazard and Occurrence Report (SHOR) or Air Safety Incident Report (ASIR);

b. Paper-based report using Air Operations - Safety Hazard & Occurrence Report (A-SHOR) form available on board the aircraft or in every operation office at outstation then send the filled form to the SS Directorate office or via email;

c. Report to SS Directorate via email; or

d. Report using short message service/instant message (chat) to SS Directorate number. The contain of the report must follow the following format:

```
A. Aircraft     : registration and aircraft type
B. Date         : date of occurrence
O. Occurrence   : chronology of occurrence
R. Route        : number and flight route
E. Evidence     : evidence of occurrence
R.              : name of reporter
```

If the report content received by the SS Department is not sufficient as a source to initiate an investigation, the reporter will be called back to complete the report.

SMSM subchapter 3.5.1 "Reception of Reports" described the process after safety report is received by the SS Department, in part, as follows:

Upon receiving the reports, safety department will validate and review the report to determine whether a real safety issue exists and make a justification whether or not the report requires an imminent action and need to be resolved immediately. Feedback will be sent to the reporter as acknowledgement.

If the report is classified as subject to mandatory reporting to the authority, initial notification shall be made refer to CASR Part 830.

The Safety and Security Department personnel who validate and review safety report worked following normal office hour (Monday to Friday, from 0830 to 1730 LT).

According to the SMSM subchapter 5.2, the Lion Air has risk management strategies comprises reactive, proactive and predictive strategies. The proactive strategies included hazard and incident reporting systems to promote identification of latent unsafe conditions.

The hazard identification process described in the SMSM subchapter 5.3 by providing examples of hazard as follows:

• Design factors, including equipment and task design;

• *Procedures and operating practices, including their documentation and checklists, and their validation under actual operating conditions;*

• *Communications, including the medium, terminology and language;*

• *Personnel factors, such as company policies for recruitment, training and remuneration;*

• *Organizational factors, such as the compatibility of production and safety goals, the allocation of resources, operating pressures and the corporate safety culture;*

• *Work environment factors, such as ambient noise and vibration, temperature, lighting and the availability of protective equipment and clothing;*

• *Regulatory oversight factors, including the applicability and enforceability of regulations; the certification of equipment, personnel and procedures; and the adequacy of surveillance audits; and*

• *Defenses, including such factors as the provision of adequate detection and warning system, the error tolerance.*

In regards to ensure Lion Air personnel are trained and competent to perform the SMS duties as part of safety promotion activities, the SMSM subchapter 7.4 described SMS training is mandatory for all operational personnel. The SMS initial training is conducted in a period of 8 hours.

The initial SMS training material revision 00-2019, described the new version web-based application software which allowed reporter to fill online form of Safety/Hazard Observation Report (SHOR) or Air Safety Report (ASR).

Operational personnel must conduct SMS recurrent training periodically not more than 36 months. The SMS recurrent training is conducted in a period of 2 hours.

The investigation took sample of hazard report record to get better understanding of the hazard reporting system. On December 2018, the SS Department received 930 safety reports which consisted of 51 hazard reports and 879 occurrence reports. The occurrence report generally contain report of go around, delay, defect to the aircraft including aircraft furnishing and unruly passenger.

Lion Air Maintenance Management

The Lion Air maintenance management was the responsibility of the Maintenance and Engineering Directorate lead by a Director of Maintenance and Engineering. The maintenance performances of all Lion Air fleet are subcontracted to an Approved Maintenance Organization (AMO). The monitoring of the daily activity was conducted by the Quality Control division, Planning and Technical Services division, and Flight Maintenance Management division that are responsible to Maintenance and Engineering Director.

(a) Lion air is responsible for the airworthiness of its aircraft, including airframe, aircraft engines, appliances, and related parts; the performance of the maintenance, preventive maintenance, and alteration of its aircraft, including airframes, aircraft engines, appliances, emergency equipment, and related parts in accordance with the CMM and the CASR Part 43.

(b) Although Lion Air may make arrangements with another person for the performance of any maintenance, preventive maintenance or alteration, however, this does not relieve Lion Air of the responsibility specified in paragraph 121.363 (a).

Maintenance Performance

The Lion Air eligible to conduct the aircraft maintenance with the scope as stated in the Authorization, Conditions and Limitation (ACL) approved by the DGCA. The investigation collected the latest revision of the ACL which was approved by DGCA on 4 October 2018. The ACL permitted the Lion Air to subcontract the maintenance performance to another entity.

The subcontracted AMO should conduct the maintenance performance base on the inspection program set forth in Company

Maintenance Manual (CMM) (issued 06, revision 00, date 27 June 2018) chapter 3 as follows:

INSPECTION PROGRAM

Inspection

General

- *This chapter describes the basic inspection policies and procedures governing the operations of Lion Air policies and procedures to suit airline's organization and Indonesia DGCA requirement ref. to CASR part 121.367.*
- *Lion Air authorized by its Operations Specification to operate airplanes under an Approved Maintenance Program (MP). To ensure that airworthiness of an operated aircraft is properly maintained, the individual Approved maintenance program (MP) shall be used and implemented.*
- *Lion Air Maintenance & Engineering Quality Control Division is responsible for ensuring that the aircrafts operated are complying with standard airworthiness, and fulfill Lion Air requirements.*
- *The policies and procedures set forth in this chapter are intended to guide Quality Control personnel in the proper fulfillment of their responsibilities.*

Method of compliance

- *All maintenance, preventive maintenance, alteration of the aircraft, engines, components and appliances for continued airworthiness are carried out by a contracted and Approved Maintenance Organization, that shall be performed in accordance with the method techniques as prescribed in the current manufacturers Maintenance Manuals, Lion Air MP, policy and procedure or other methods techniques, and practices acceptable to the DGCA.*
- *Subcontractor's tools equipment and test apparatus necessary to assure completion of the work in accordance with accepted industry practices shall be used. Where special equipment or test apparatus is recommended by the manufacturer involved, that equipment shall be used or apparatus or its equivalent if acceptable to the Director General must be use.*
- *Each person performing maintenance, preventive maintenance and alteration shall do that work in such manner and use material of such quality that the condition of the aircrafts, engines, appliances work on shall be in conformity to Manufacturer approved definition and approved maintenance procedures (Reference to Manufacturer IPC, AMM, and SRM).*

Refer to the ACL that approved on 4 October 2018, the applicable Continuous Airworthiness Maintenance Program (CAMP) for Boeing 737-8 (MAX) referred to document number LA-CM-02-08, issue 02, revision

00, dated 28 February 2018. The PK-LQP aircraft was not include in the aircraft applicability page of the CAMP. PK-LQP listed in the ACL since June 2018.

The Temporary Revision of the CAMP identified as document number 001/B737MAX/CAMP/VII/2018 dated 13 July 2018 to amend the aircraft list to include the PK-LQP. The temporary revision document was submitted to the DGCA on 29 August 2018. The investigation did not receive the ACL revision to reflect the CAMP temporary revision.

The maintenance performances of all Lion Air fleet are subcontracted to an AMO which was performed by Batam Aero Technic (BAT) under the contractual agreement number 039/JT/DE/XI/2014 Amendment III or 010/BAT/Agreement/JT/XI/2014 Amendment III referred as Total Maintenance Care.

The Total Maintenance Care agreement covered all Lion Air aircraft including the Boeing 737-8 (MAX) aircraft has the scope of work as follows:

- Scheduled line and base maintenance as specified in the approved Lion Air CAMP or other instructions for cabin maintenance and in respect of Lion Air policy for passenger treatment.
- Unscheduled maintenance performance (line and base maintenance findings).
- All deferred defect management and correction to the first opportunity.
- Airworthiness Directives (AD) implementation according to Lion Air schedule planning.
- All structure repairs of aircraft according to an approved definition.

The Total Maintenance Care agreement mentioned the Continuous Airworthiness Management agreement (identified as agreement number 038/JT/DE/XI/2014 or 009/BAT/Agreement/JT/XI/2014) which covered the engineering task for Lion Air aircraft that are maintained by BAT.

The Lion Air Quality Control department responsible to conduct the supervision of the inspection activities delegated to the BAT at the maintenance base and all line station facilities. The Lion Air Quality Control shall ensure that Lion Air aircraft and maintenance procedures comply with regulatory rules and Lion Air standard concerning airworthiness and safety (Quality Control Procedure Manual chapter 2.3.1 Quality Control Manager Responsibility).

The CMM chapter 2.2.4.1: Responsibility stated that the Fleet Maintenance Management (FMM) Manager is responsible to control and monitor the repetitive discrepancies and shall coordinate with the BAT Maintenance Control Centre (MCC) include:

• Control and monitoring of daily aircraft status issued by the BAT MCC for Lion Air aircraft operational Status (Aircraft on Ground, serviceable or phase check),

• Daily check of the BAT MCC aircraft technical status and control for repetitive discrepancies to be identified and treated by the BAT MCC as Deferred Maintenance Item (DMI),

• Control and monitor technical dispatch reliability, deferred maintenance activities, and aircraft appearance including repetitive problem correction, navigation data base of the flight management computer update.

AFML

The description of the maintenance log in chapter 4 of the CMM was as follows:

Maintenance Log
General
Discrepancies occurring during line operations of airplanes shall be recorded in the Lion Air Airplane Log Book System. The Log Book System consists of 5 (Five) basic bound volumes carried on board all Lion Air airplanes, they are:
1. Aircraft Flight & Maintenance Log (AFML)
2. Cabin Maintenance Log (CML) 127
3. Deferred Maintenance Items (DMI)
4. Non-Safety Related Deferred Item (NSRDI)
5. Dent Buckle Chart Supplement (DBCS)

Aircraft Flight and Maintenance Log (AFML) Book
General

CASR 121.709 provides that each person who takes action in the case of a reported or observed failure or malfunction of an airframe, engine, or appliance that is critical to the safety of flight shall make, or have made, a record of that action in the airplane's AFML. Therefore, the following procedure shall apply:
1. Unscheduled Maintenance performed on in service aircraft shall be documented in the Aircraft Flight and Maintenance Log book. This would apply to discrepancies observed during thru flight.
2. Additionally, scheduled maintenance performed on in service aircraft that is critical to the safety of flight (i.e., engine change, flight control replacement, etc.) shall be

documented in the Aircraft Flight and Maintenance Log book by which the flight crew would be aware of the significant maintenance accomplished on in service aircraft.
3. The following procedure addresses the entries required by maintenance personnel.

Aircraft Flight and Maintenance Log entry procedure:

a. The Captain is responsible for completing the Aircraft Flight and Maintenance Log entry. However, he may delegate this responsibility to another crew member.
b. Prior to the flight, Flight Crew shall verify that the AFML is on board in the aircraft and that it contains a sufficient number of pages for the flight scheduled.
c. The flight crew will review the AFML for corrective action taken on previous flight irregularities, type of service performed, and maintenance release, when required.
d. Discrepancies shall be recorded in the AFML. Verbal reporting to maintenance personnel may be required to add information related to the discrepancies recorded in the AFML.
e. Pilot report entry shall contain sufficient detail to assist maintenance personnel conducting the necessary corrective action, which includes the De-Icing application at the Airport having icing condition as described at OM Chapter 8.1.4.5 Landing Limitations. Any flight crew entry must be signed by the Captain.
f. Any time maintenance personnel make an entry it must be signed by the person making the entry.
g. Maintenance personnel are responsible for assuring that any log entries accurate and completed. Upon completion of a maintenance check, enter date, signature, and authorization number in the block titled "Return to Service" column. Prior to departure, Maintenance Release sign shall be entered including the required data on the column located in the lower right hand corner of the AFML.
h. All entries shall be printed in black or blue ink and must be legible.
i. Engineer's normal signature will be entered in legible manner.
j. If an error is made, a line will be drawn through the entry and new entry made or the word "VOID" is written across the log sheet, not allowed to erase an error entry, and error entry shall remain readable. Write the name and signature of the person who void the entry.
k. Do not remove/destroy/detached an error or a void log page, because DGCA requires that all log pages should be accounted for.
m. Maintenance Action entries will list any work done to correct deferred Item, or describes troubleshooting accomplished to correct a malfunction or pilots report. Such statements are necessary for records purposes and to eliminate repeat the work unnecessarily. The person making the corrective action entry is required to ensure that all works, checks, or inspections were performed in an appropriate manner and in accordance with the approve manuals and procedures.
n. CASR requires a complete description of the corrective action taken to correct a discrepancy and release an aircraft in airworthy status. Therefore use of words such as "Repaired"; "Fixed"; or "Corrected" as the sole entry for corrective action is not acceptable. The log book entry should also include a description of the troubleshooting

procedure and/or reference to the manual system that was used to correct the discrepancy.

o. Occasionally a reported malfunction cannot be duplicated, identified, or corrected (one-time occurrence). In such cases, an appropriate action taken entry will be:
• *Installation inspected, functionally checked, and found satisfactory.*
• *Request further observation and information next flight.*
p. In no case shall the word "temporary" be used to describe the extent of a repair, but rather the word "airworthy" should be used.

Responsibility

An Aircraft Technical Log is required to be used by all operators approved in accordance with CASR 121.

According to the contract between 121 Lion Air M&E and BAT; Lion Air request AMO subcontractors to fill by Authorized personnel's and to record: Each schedule and unscheduled maintenance operation performed between two Base Maintenance Inspection including each malfunction detected by the crew in operations.

The AFML is a major document for Airworthiness Control, including following key information:

- *The Line Maintenance schedule maintenance performed in accordance with the Maintenance Program (MP) (For example: Pre-flight check, daily check, service check, engine oil consumptions, etc.)*
- *All deferred maintenance items and maintenance corrective actions,*
- *Rotable changes,*
- *Certificate of release to service and Maintenance Release,*
- *All structural defect arising during operation.*

According contract, BAT AMO shall review and check on daily basis all AFML for Lion Air fleet:

- *Proper filling of the document by all users according to 121 Lion Air procedures,*
- *Sequences number of AFML should be checked for completion,*
- *Proper integration and control of all AFML information's in the*

  Maintenance Information System (Aircraft Hours, Cycles, Scheduled Maintenance tasks, malfunctions and all unscheduled Maintenance tasks, Deferred defects, fluid consumption, etc.)
- *Analyze performed Line Maintenance activities to fulfill Maintenance*

Program (MP) requirements and applicable regulation (Ex: Pre-flight Check, Daily Check, Service Check, etc.)

Control of the schedule line maintenance and Maintenance release certificates,

- *Control of Cabin Maintenance Log for proper utilization,*
- *The proper management of DMI and the respect of MEL items intervals.*

According contract BAT AMO shall sort and archive AFML in accordance with a proper archive plan the original Aircraft Technical Log in a protected and controlled area and provide access to 121 Lion Air, Authorities representatives, owner or lessor upon request. 121 Lion document revision a direct presentation of the revised documents to Pilots Manager and to AMO Line Maintenance and Quality Manager is organized for AFML Socialization.

FTIM Lion Air Manual is dispatched and always available to BAT AMO staffs. Lion Air M&E, Quality Control Division, is controlling the AFML format (FTIM) and utilization by Lion Air Pilots and AMO Line Maintenance Technicians.

Refer to CMM Chapter 4.3 the AFML should be archived and saved properly. The CMM procedure for the record handling is shown below.

MAINTENANCE RECORD AND RETENTION

Technical Record

Introduction
This chapter describes a Technical Records control system in Lion Air. The retention of records enables a Lion Air to maximize the value and utilization of its assets (airplanes, engines, components, appliances, etc.) by documenting proper parts removal & installation, schedule and unscheduled maintenance task, alteration, repair, modification, maintain a list of current alterations to each airframe, engine, and appliance, the current status of life limited parts of each airframe, engine, and appliance. The chapter describes the policies and procedures to be used by Technical Records, under the supervision of Planning & Technical Services Manager and control of the Quality Control Manager. The policies and procedures are designed to meet the requirements of CASR 121.369 and 121.380.

General
Technical Records will be filed and stored in such a manner that the records will be readily accessible to authorized auditors and other personnel. The filing and storage system shall promote easy of retrieval, accessibility and adequate controls.

Filling Provisions

a. Recording Maintenance Data

1) Recording maintenance data means to record all maintenance; preventive maintenance and alteration activities inspection perform and up to approved Maintenance Release is signed off as required by CASR 121.709.

2) All airplane records are included in the part of the Aircraft Flight and Maintenance Log (AFML), which should meet as required under CASR 121.380 and 121.709 otherwise causes to be made maintenance records falsification as stated in CASR 43.12.

3) The recording of all work carried out on airplane, components and issuance of instruction records, receipts, certificates, etc. shall be accomplished by the used of appropriate approved forms.

4) All forms that record details of overhaul, repair or modification of an airplane or component must be processed and retained as part of current and complete history of the aircraft, or component, in a condition presentable to Indonesian DGCA.

5) All maintenance personnel are responsible for accurate recording and documenting of work performed and approved, using an appropriate company document to determine its compliance with the Indonesian CASR's.

6) Base on CASR 121 380 (c) all maintenance records required to be kept by this section available for inspection by the Director

b. Handling Technical Records

Technical records handling Means a handling all of administrative maintenance document and technical data respecting to the maintenance, preventive maintenance and alteration work performed. These records are filed and kept in the Technical Record subdivision.

c. Filing of Maintenance Document

Filing maintenance document and technical data means retrievable filing of all maintenance documents and technical data for which a retention period as required under CASR 121.380, 120.380a and 121.709.

A copy of each Job Order with all attached supplementary form shall be maintained in the Planning & Technical Services.

A Planning & Technical Services ensure each maintenance document for work accountability, parts used; engineers; and Inspectors sign or authorized personnel who certify a maintenance release before such document send to technical record for proper filing.

The investigation received AFML record on October 2018 of PK-LQP. The AFML page can be identified by the page number on the top right of the page, such as B1800499, B3042854, B3042855, etc. The investigation found 31 pages not included in the package.

Lion Air Repetitive Problem Handling

The Lion Air referred the repetitive problem as repeat discrepancy. The repeat discrepancy handling is described in the chapter 14.2: Repeat Discrepancy Control System, which described as follows:

Repeat Discrepancy Control System

General
a. The purpose of the Repeat Discrepancy Control System is to identify and correct any discrepancies that recur within a 5 (five) days period.
b. It is a Lion Air policy that any discrepancy which twice recurs on the same airplane during 30 (thirty) consecutive days of operation will be identified as a "repeat discrepancy" (refer CMM Chap. 15.2.3. As soon identify FMM Division will control the registration by AMO subcontractors on DMI list of this repeat discrepancy. A positive plan of corrective action is pursued to preclude further recurrence of such discrepancies and aircraft shall not be returned to service unless the root cause is affirmatively corrected.
c. Every effort shall be made to prevent repeat discrepancies in order to eliminate unnecessary airplane out of service time and the cost unnecessary for component inspection and overhauls.

Responsibility
Lion Air M & E is subcontracting all maintenance to AMO subcontractors. AMO Engineer on duty is responsible for the coordination of all activities relating to repeat discrepancies occurring during line operations.

Lion Air M & E has contracted BAT AMO to provide Maintenance Control Center for all Lion Air fleet. One of MCC responsibilities is to detect all repeat discrepancies and to define corrective actions.

a. The MCC Repeat Analyst is controlled by Lion Air M & E FMM which is the local point for implementing of the Repeat Discrepancy Control System.
b. The repeat Discrepancy Control System consists of the following elements:
1) Identification of trouble and root causes
2) Post rectification or repair attempt and its result
3) Review and follow up
4) Surveillance
FMM Manager is reporting to regularly to Lion Air M & E director for all repeat discrepancy issues.

Procedure

1) Identification
a. The Repeat Analyst will be monitored by continuously reviewed the maintenance records of the previous 30 (thirty) days. The review will include:

i. Corrected and uncorrected pilot complain, and ground write-ups from the computer databases for each airplane.
ii. Deferred maintenance items recorded and tracked by FMM
iii. Daily activity logs recorded by FMM. The logs are the records of discrepancies reported by flight crew and line stations.
iv. Daily maintenance advises at home base and out stations.
b. When an Item has been identified as a repeat discrepancy, the Repeat Analyst will coordinate its rectification, as follows:
i. If the repeat discrepancy can be rectified during line maintenance at the home base line stations or at an out-station, the Repeat Analyst coordinates the rectification with FMM
ii. If the repeat discrepancy can only be rectified during home base hangar maintenance, the Repeat Analyst will coordinate rectification with Planning & Technical Services Division.
c. All item identified as repeat discrepancies will be brought to the attention of the Fleet Maintenance Management and Quality Control Manager on a daily basis.
d. The Repeat Analyst and Lion Air FMM Manager will maintain a log of all identified repeat discrepancies for tracking and follow up purposes.

2. Rectification
Lion Air have a documentation all troubleshooting history and method rectification that used in previous repair attempts for tracking chronic or repetitive discrepancies aircraft experiences.

a. The Repeat Analyst will generate the appropriate repeat discrepancy work forms for each repeat discrepancy identified:

1) The repeat discrepancy assigned to home base hangar maintenance for rectification will be recorded on a Non-Routine Write form.
2) The repeat discrepancy assigned to line stations (including home base line maintenance) for rectification will be recorded in a Deferred Maintenance Item form. (ref. FTIM LA-TF-03-01)

b. For repeat discrepancies rectified at home base AMO hangar maintenance, AMO subcontractor will make the necessary corrective action for the repair: AMO Subcontractor will package the repeat discrepancy work form together with a specific check for accomplishment under Lion Air PTS approval

c. For repeat discrepancies which be rectified at a line station (including the home base ine station), MCC BAT on duty will make the necessary arrangement for the repair:

1) The AMO line station maintenance personnel assigned to work on the airplane will be notified of the repeat discrepancy and its history.

2) Additional expertise may be requested to assign to assist in trouble shooting the repeat discrepancy.
3) Rectification of the repeat discrepancy will be recorded and signed off on the AFML book in a manner similar to Deferred Maintenance Item.
4) All returned repeat discrepancy work cards and/or maintenance log pages will be forwarded to the Repeat Analyst for review.

d. If the repeat discrepancy deferrable and troubleshooting/repair of the item will exceed scheduled ground time:

1) The FMM will deferral authority from the line maintenance.
2) Based on the severity or operational impact of the discrepancy, the FMM may suggest the airplane be removed from service to facilitate corrective action.
3) If repair of the discrepancy is deferred to a subsequent line maintenance opportunity, FMM will continue to track the discrepancy and will make arrangement to have trouble shooting/rectification accomplished on the rectification accomplished on the item at the first opportunity.
4) If repair can be deferred to the next home base hangar maintenance opportunity, responsibility for the rectification, together with the appropriate paper work, will be transferred to Planning & Technical Services Division.

e. Review and Follow up

a. The Repeat Analyst will review returned work form and other work record, and update record, as necessary.
b. Copies of the repeat discrepancy work forms, which have been cleared by maintenance corrective action will be held in file by the Repeat Analysts for a minimum of 5 (five) days of airplane operation to assure no further recurrence.
c. If the Item does not recur for 5 (five) days, the Repeat Analyst may clear the item from the repeat discrepancy log and discard the work card. History of the Item will be stored in computer records.
d. If the Item does recur, the Repeat Analyst will notify the Quality Control Manager or his designee.

1. The Repeat Analyst will re-open the repeat discrepancy.
2. Quality Control through coordination with Planning & Technical Services will determine whether to remove the airplane from service or defer the Item.
3. Under no circumstances an airplane will be released for revenue service if an Item is found to have repeated continuously without the express authorization from Quality Control Manager.
4. Surveillance
a. The Repeat Analyst accomplishes continuing analysis and surveillance of repeat discrepancies.
b. The FMM is responsible for monitoring Items identified as repeat discrepancies.

The repetitive problem is part of the continuing analysis and surveillance that described in the CMM chapter 15 as follows:

CONTINUING ANALYSIS AND SURVEILLANCE

Introduction
This section provides a description of, and the responsibilities related to operator Continuing Analysis and Surveillance Program, as required by CASR part 121.373. Today Lion Air is organized with 2 sections working together for the performance of the DGCA and CASR requirements for the Maintenance Continuing Analysis and Surveillance Program:

• *A Quality Control Division under Maintenance & Engineering Director,*
• *A Quality Assurance Directorate, under Accountable Manager, Lion Air Maintenance and Engineering, Quality Control Division will continue to perform a part of the "Continuing Surveillance Program".*
Quality Assurance Directorate will ensure all required internal and external Quality Audits required to monitoring continuously the CASR requirements compliance.

The repetitive discrepancy criterion which was stated in the CMM chapter was described as follow.

Compliance, Conformity and Adherence to the Regulations
Quality Assurance Directorate is closely and continuously coordinating his Quality surveillance program with Quality Control Division to prevent the implementation of improvements to the aircraft damage/problems protracted, refers DGCA safety Circular No.: AU/0649/DSKU/03/2007.

Lion Air are required to:

a. Correctly and consistently implementing all the requirements outlined in the CASR 121.373, as follows:

- Internal audit performed
- To implemented an analysis of the effectiveness of maintenance programs that approved
- To carry out maintenance review board meeting to repetitive trouble
- To take notes of repetitive trouble that cannot be repaired in DMI
b. Restrict for repetitive trouble, as follows:
- The maximum occurs twice within a period of 30 days and put in DMI process, the aircraft components included in the ATA code: 21, 22, 24, 27, 28, 29, 31, 32, 34, 36, 45, 49, 52, 53, 54, 55, 56, 57, 71, 72, 73, 74, 75, 76, 77, 79, 80.
- The maximum occurs four time within a period of 30 days for aircraft components other than the ATA code above mentioned

Approved Maintenance Organization

Batam Aero Technic (BAT) Management

The Batam Aero Technic (BAT) is an Approved Maintenance Organization (AMO) under CASR 145 with the approval number 145D-914. The capability list approved by the DGCA included the maintenance activities for the all Lion Air aircraft in the base maintenance and line maintenance activities. BAT is a wholly owned subsidiary of the Lion Air Group.

The BAT established procedure manual for conducting the maintenance activity, which referred as the Batam Aero Technic and Quality System Manual (AMOQSM) with the document number BT-GEN-01. The manual described the housing, facilities, equipment, personnel and general operating rules.

The BAT has hangar facilities to provide the base maintenance activity at Surabaya, Batam, Cirebon, Palangkaraya and Balaraja.

The line maintenance activity of the Lion Air aircraft was conducted by BAT in the several transit stations including in the Jakarta, Denpasar and Manado.

The total number of engineers at Jakarta consist of about 300 personnel including 25 engineers rated for Boeing 737-8 (MAX) aircraft including authorization to release to service. The engineers divided into four shift groups. The daily maintenance activities were conducted in two shifts. The first shift started from 0700 to 1930 LT and the second shift started from 1900 to 0730 LT. The BAT handled approximately 458 flight daily at Jakarta.

The total number of engineers at Manado were 34 personnel divided into four shift groups. The maintenance activities were conducted in two shifts. The first shift started from 0700 to 1900 LT and the second shift started from 1900 to 0700 LT. The BAT handled approximately 12 – 18 aircraft transit and 10 aircraft overnight daily in Manado. The Boeing 737-8 (MAX) maintenance activity was conducted by the engineer on duty from Jakarta as no engineer in Manado was rated for Boeing 737-8 (MAX).

The total number of engineers at Denpasar were 80 personnel divided into four shift groups. The maintenance activity was conducted in two shifts. The first shift started from 0645 to 1915 LT and the second shift started from 1845 to 0715 LT. The BAT handled approximately 60 aircraft transit and 5 aircraft over night, daily in Denpasar. The Denpasar had 4 engineers rated for Boeing 737-8 (MAX) aircraft including authorized to release to service.

The BAT has Maintenance Control Center (MCC) department is led by a General Manager which responsible for the maintenance activity control. The job title of General Manager MCC was stated in the AMOQSM but the detail of the duty and responsibility of the MCC was not described.

The Lion Air and BAT agreed to monitor the aircraft serviceability through dedicated group personnel referred as Bapak Asuh lead by Manager Bapak Asuh. The job title Manager Bapak Asuh stated in the AMOQSM however, the duty and responsibility of Manager Bapak Asuh was not described.

The maintenance performance for the customer work performance procedure as follows:

WORK PERFORMED PROCEDURE FOR CERTAIN OPERATORS

- *Reference: CASR Part 145.205 Maintenance, preventive maintenance, and alterations performed for certificate holders under parts 121, 135, and for foreign air carriers or foreign persons operating an Indonesian registered aircraft in common carriage under CASR part 129, CASR Part 209 AMO (Approved Maintenance Organization) manual content.*

Maintenance work to be performed by the Batam Aero Technic for the organization that has a continuous airworthiness maintenance program under the certificate part 121 and 135 or 129 shall be follow the operator's maintenance program.
In addition, Batam Aero Technic also perform maintenance on operators that have an approved maintenance program by authority under section 141 certificates.
The President Director through the General Manager Quality and Engineering Manager establishes an understanding of the contractual arrangements for the particular requirements of each customer.

The customer's requirements may include but are not limited to:

- *Knowledge and/or training requirements for technicians and inspectors assigned to perform work;*
- *Review and acceptance of documents used by the Batam Aero Technic when recording work performed;*
- *Following specified procedures contained in the customer's operations specifications, maintenance program or as set forth in the general maintenance manual; and,*

• *Specific approvals or authority to perform major repairs or alterations.*
If the customer provides technical data, it will be referenced in the applicable maintenance record. If there is any question as to what technical data is to be used at any time during performance of work, it will be brought to the attention of management or the customer for appropriate resolution. The Customer Specific Requirements list sets forth the special instructions from each air carrier customer and is used as a reference guide to ensure proper maintenance is performed as required by the air carrier or commercial operators' manual.

In addition, through audits by the air carriers and commercial operators, this Batam Aero Technic will ensure continuation compliances with the customer's requirements with respect to record keeping, training of personnel and other matters covered by the customer's contract and/or maintenance manual procedures.

Defect Handling

As part of the Total Care maintenance agreement, the BAT engineers are responsible to perform the aircraft trouble rectification. The BAT Line Maintenance Procedure Manual (LMPM) describes the procedure for line maintenance activity including the defect rectification and administers the AFML page. The descriptions of the LMPM regarding the defect rectification are shown below.

MAINTENANCE DEFECT RECORD
Maintenance Defect Identification
This procedure explains how to perform the AFML record associated to the procedure LMPM

"AIRCRAFT MAINTENANCE DEFECT HANDLING".
During the course of operations, a defect can be identified and recorded
• *By the PIC: in this case the PIC is responsible to report the defect through a PIREPS as per their operator procedures (discrepancies block), or*
• *By the engineer (preflight, transit, daily or service check, out of phase maintenance, special inspections, etc.): in this case the individual is responsible to record the defect in the discrepancies block and to tick the "MA" block (Maintenance Action block)*
• *When a defect is mentioned by crew or maintenance team, the engineer shall check on previous technical log book pages if the same problem has already occured, to identified as repetitive problem.*
• *Defect that identified as repetitive problem should be notified to Specialist department and Bapak Asuh for further deep monitoring, evaluation and troubleshooting.*
• *Problem occurrences more than 3 (three) times within 30 days maintenance should be deferred on DMI and report to MCC.*
Maintenance Defect rectification

This procedure explains how to perform the AFML record associated to the procedures LMPM 3.5 "AIRCRAFT MAINTENANCE DEFECT RECTIFICATION".
The defect rectification is recorded by the engineer in the "action taken block".
The following information must be recorded in the AFML:
- *Details of maintenance performed,*
- *Reference to the approved data (page / block / revision date),*
- *Reference to removed / installed component (PN / SN),*
- *ATA code (6 digits),*
- *If applicable, reference to the Work Order, Task Card, EO or EA,*
- *If applicable, dimension or test figures within a specific tolerance, the exact dimension or test figures shall be recorded (it is not normally sufficient to state that the dimension or the test figure is within tolerance),*
- *If applicable, detail of RII or duplicate inspection,*
- *Return to Service information.*

The LMPM described in the Maintenance Defect rectification above written as "AIRCRAFT MAINTENANCE DEFECT RECTIFICATION", however the chapter in the LMPM was "ENGINE RUN UP PERFORMANCE".

On-Board Maintenance Function (OMF) Handling

The Boeing 737-8 (MAX) equipped with the downloadable On-Board Maintenance Function (OMF) which provide the historical record of the aircraft problem.

The investigation found that the OMF data download, referred to the task card 737M-46-INT-00-01-MLI Revision 01 was not included in the Boeing 737-8 (MAX) CAMP.

Refer to the PK-LQP AFML, the last OMF data download was performed on 25 October 2018 at Denpasar.

The investigation collected the OMF data of PK-LQP from the Flight Data Service department of the BAT Engineering and DOA division. The duties and responsibility of the Assistant Manager of Flight Data Services was described in chapter 2.2.2.5 in the Engineering Procedure Manual. The Engineering Procedure Manual was not mentioned that the OMF data management as the duties and responsibility of the Assistant Manager of Flight Data Services.

The investigation did not find any manual describes the policy and procedure of handling the OMF.

BAT Repetitive Problem Handling

The BAT defined aircraft problem categorized as repetitive (repeat discrepancy) if a same problem in the same aircraft registration occurred repeatedly minimum 3 times within 30 days. The aircraft repetitive problem was handled by the Manager Specialist in coordination with the *Bapak Asuh* and other departments as stated in the LMPM chapter 2.2.2. The duties and responsibilities of the Manager Specialist are as follows:

MANAGER SPECIALIST
Duty & Responsibilities
• *Responsible to make a report to Line Maintenance General Manager*
• *Ensure his subordinates to monitor Repetitive problem by Trax, PFR (Post Flight Report), etc.*
• *Ensure his subordinates to follow up Repetitive problem, and guidance for trouble shooting.*
• *Ensure to analyze for reducing repetitive problem by collect data repetitive, look up to Manufacture Document and action (SL and SB) – Note: Service Letter and Service Bulletin.*
• *Ensure for part needed and part improvement (upgrade due to low reliability of Part) to engineering and Material Planning.*
• *Give guidance for his/ her subordinates.*
• *Coordinates with other department such as Line Maintenance, Bapak Asuh, PPC, material planning, etc in order to make immediate rectification about repetitive defect on aircraft.*

The repeat discrepancy handling is described in the LMPM chapter 13.2: Control of Repeat Discrepancies, as shown below.

CONTROL OF REPEAT DISCREPANCIES

PURPOSE
The purpose of this procedure is to monitor and control the repeat/ recurring discrepancies to the aircraft in order to initiate special actions to preclude further recurrence.
This procedure is applicable to customer being maintained by Batam Aero Technic under the sub contracted maintenance agreement.

PROCEDURE
• *Repetitive discrepancy is defined as a same problem in the same aircraft registration occurred repeatedly minimum 3 times within 30 days. will be identified as a "repeat discrepancy" and to be recorded in DMI Log.*

- *Specialist team support is assigned to monitor and maintaining oversight over discrepancies reported in operating aircraft and identifying the repeat discrepancies by TRAX*
- *Specialist team in coordination with the involved department, reviews maintenance records of the previous day for each aircraft in order to identify repetitive discrepancies of similar nature, on same ATA Chapter and Section, affecting to same equipment or for analogous causes. The reviews include:*
  - *Corrected and uncorrected aircraft discrepancies.*
  - *Deferred maintenance items recorded.*
  - *Daily activity book record discrepancy reported by flight crews and line maintenance personnel at main base and outstations.*
  - *Discrepancies raised in the operator log book system.*
  - *DMI Listing Summary List*
  - *MDRR*
- *When an item is identified as repetitive defect that cannot be solve in the Line Maintenance, Specialist team will coordinate a meeting with all the concerned departments (example; QC inspectors, reliability engineers, etc.) and representatives of operator/ customer in order to discuss the repeated discrepancies, identify its root causes and propose a corrective action plan. Minutes of the meeting as well as the proposed corrective action plan must be maintained.*
- *Specialist can raise a "Work order" or a "Job card" liked to the DMI on TRAX, if the engineer is not able to solve the problem without technical assistance.*
- *Corrective action plan is issued in a work order identified as "repeat discrepancy" and scheduled to be complied as soon as possible in accordance with significance of the defect.*
- *Specialist coordinates the rectification in accordance with the agreed corrective action plan starting the process of rectification and monitoring its progress until final closure.*
- *A database of repetitive defects per aircraft type is maintained TRAX by specialist or by Bapak Asuh.*

The BAT LMPM contains the policy to handle repeat/repetitive discrepancies (repetitive problem) but did not contain detail procedure.

Regarding the repetitive problem handling of SPD and ALT flags that occurred on PK-LQP since 26 October 2018 until the last record, the investigation did not find any document or discussion regarding the handling of this repetitive problem. The investigation only found the statement of repetitive problem which was written by Denpasar engineer in the AFML when performing the problem rectification at Denpasar on 28 October 2018.

Xtra Aerospace LLC

Xtra Aerospace LLC (Xtra Aerospace) has an office and facility in Miramar, Florida and it is an Approved Maintenance Organization (AMO) that was approved by the Federal Aviation Administration (FAA). The FAA Miramar Flight Standards District Office (FSDO) issued approval as

Repair Station Certificate to Xtra Aerospace. This entitled Xtra Aerospace to conduct the maintenance, repair and overhaul for limited accessories, limited instruments, limited radio and emergency equipment as stipulated in the operations specifications.

As part of the FAA approval, and when the accident aircraft AOA sensor was repaired, Xtra Aerospace had the approved manuals of Repair Station Manual (RSM) and Quality Control Manual (QCM). The Director of Quality is responsible for keeping the Work Order Package which contains the customer commercial invoice, repair order, packing list, and proforma invoice including authorized release certificate (FAA form 8130-3). Work order teardown report, component traveler, test data sheet, and packing slip.

The Angle of Attack (AOA) sensor part number 0861FL1 (manufactured by the Rosemount Aerospace as Original Equipment Manufacturer – OEM) was, at the time of the Lion Air Accident, one of the capabilities that Xtra Aerospace was approved to perform. Shortly after the accident, Xtra Aerospace discontinued any further work on the 0861FL1 AOA sensors.

The RSM chapter 2 describes that the engineering department is responsible to ensure equivalencies of tool and equipment including evaluation of equivalency, and maintaining pertinent documentation.
The requirement for assessing the equivalency of test equipment is described in the chapter 5 of RSM as shown below.

Equivalent Test Equipment

The "Equivalent Test Equipment" must be capable of performing all normal tests and checking all parameters of the equipment under test, ARINC 668 "GUIDANCE

FOR TOOL AND TEST EQUIPMENT EQUIVALENCY" *shall be used for guidance during this process. The Equivalence Determination Procedure shall be documented using XTA Form 181. (A sample of XTA Form 181 is located in the Quality Publications Appendix D).*

- *The level of accuracy should be equal or better than that recommended by the Original Equipment Manufacturer (OEM).*
- *OEM technical data shall be used in the manufacturing of "Equivalent Test Equipment".*

- OEM *calibration or inspection procedures shall be used if available for the certification of the "Equivalent Test Equipment". A certification procedure shall be developed for equipment on which the OEM does not support.*
- *"Equivalent Test Equipment" shall meet the same requirements for identification and control.*

The Rosemount Aerospace (now part of Collins Aerospace) Component Maintenance Manual (CMM) 34-12-34 revision 8 recommends the test and calibration equipment for repair of the AOA sensor with part number of 0861DR, 0861FL and 0861FL1 is the North Atlantic 8810 (later versions recommend the 8810A). The CMM (34-12-34) does have a note that states equivalent substitutes may be used.

Xtra Aerospace may utilize the Peak Electronics SRI-201B (Model 7724-00-2) for test and calibration of the AOA sensor with part number 0861DR, 0861FL and 0861FL1. The test equipment equivalency certificate was provided for the Peak electronics API SRI-201B (model 7724-00-2). The Test Equipment Equivalency listed the Peak Electronics SRI-201B and justified the equivalency as "specification comparison." Xtra Aerospace did not develop written instructions for the use of Peak Electronics SRI-201B (Model 7724-00-2) during the testing. CMM 34-12-34 also does not specify specific written instructions for the North Atlantic 8810.

The Quality Control Manual stated that the test equipment shall be calibrated at periodic intervals established on the basis of stability, purpose and degree of usage as shown in the chapter 6 of Repair Station Manual as shown below.

Test Equipment Calibration Requirements

Test equipment shall be calibrated at periodic intervals established on the basis of stability, purpose and degree of usage. One year shall be the standard interval unless the manufacturer of the equipment has authorized a longer calibration interval. Calibration is valid until the last day of the month due as recorded on the calibration certificate/label. All calibrations shall be performed before or on the last day on the month of expiration.

a) All meters and measuring equipment will be outsourced to an approved metrology laboratory for calibration. All calibrated standards and equipment shall be traceable to the National Institute of Standards & Technology (NIST) or an acceptable foreign or international standard.

b) *All Complex Test Panels used to interface units with measuring equipment shall undergo in-house inspections using controlled OEM procedures or XTA certification procedures; follow-up inspection/verification will be accomplished as required.*

The equipment calibration record showed that all equipment was acceptable and within each equipment's requirements..

Persons performing maintenance (including inspections), preventive maintenance, and alteration must be assessed and trained in accordance with the FAA approved procedures set forth in Xtra Aerospace LLC. training manual.

Xtra Aerospace relied upon On-the-job training (OJT) to ensure technicians who are performing the work are fully informed about the procedures, techniques, and equipment maintained by Xtra Aerospace and to ensure each person is qualified to perform the duties. OJT is conducted at the work site by a supervisor, designated instructor, or manufacturer representative. OJT is accomplished by utilizing the applicable technical publications that contain the approved/accepted procedures for the maintenance or inspection task to be accomplished.

Minimum Equipment List Regulations

The CASR required the operator to utilize the Minimum Equipment List (MEL) as dispatch guidance when one or more instruments or components are inoperative.

The flight Instruments and equipment that are either specifically or required by the airworthiness requirements is described in the CASR 121.303.

CASR Part 121.303 Airplanes Instruments and Equipment
(b) Instruments and equipment required by Sections 121.305 through 121.359 must be approved and installed in accordance with the airworthiness requirements applicable to them.

The Indonesia regulation related to the airspeed and altitude indicator is described in the CASR 121.305 as follow.
CASR Part 121.305 Flight and Navigational Equipment

No person may operate an airplane unless it is equipped with the following flight and navigational instruments and equipment:

(a) An airspeed indicating system with heated pitot tube or equivalent means for preventing malfunctioning due to icing.
(b) Two sensitive pressure altimeters with counter drum pointer or equivalent presentation.
(c) An accurate timepiece indicating the time in hours, minutes and seconds.
(d) A free air temperature indicator.
(e) A gyroscopic bank and pitch indicator (artificial horizon).
(f) A gyroscopic rate of turn indicator combined with an integral slip/skid indicator (turn and bank indicator) except that only a slip/skid indicator is required when a third attitude instrument system usable through flight attitudes of 360 of pitch installed in accordance with Paragraph (j) of this section.
(g) A gyroscopic direction indicator (directional gyro or equivalent).
(h) A magnetic compass.
(i) A vertical speed indicator (rate of climb indicator).
(j) On the airplane described in this paragraph, in addition to two gyroscopic bank and pitch indicators (artificial horizons) for use at the pilot stations, a third such instrument is installed in accordance with paragraph (k) of this section:

(1) On each turbojet powered airplane.
(2) On each turbo propeller powered airplane having a passenger-seat configuration of more than 30 seats, excluding each crewmember seat, or a payload capacity of more than 7,500 pounds
(k) When required by Paragraph (j) of this section, a third gyroscopic bank-and-pitch indicator (artificial horizon) that:
(1) Is powered from a source independent of the electrical generating system;
(2) Continues reliable operation for a minimum of 30 minutes after total failure of the electrical generating system;
(3) Operates independently of any other attitude indicating system;
(4) Is operative without selection after total failure of the electrical generating system;
(5) Is located on the instrument panel in a position acceptable to the Director that will make it plainly visible to and usable by each pilot at his or her station; and
(6) Is appropriately lighted during all phases of operation.

The regulation required that the aircraft instruments and equipment required by an airworthiness directive shall be operable as described in the CASR 121.628 below.

CASR Part 121.628 Inoperable Instruments and Equipment
(a) No person may take off an airplane with inoperable instruments or equipment installed unless the following conditions are met:
(1) An approved Minimum Equipment List must be carried onboard the airplane
(2) The DGCA shall issue the certificate holder operations specifications authorizing operations in accordance with an approved Minimum Equipment List. The flight crew shall have direct access at all times prior to flight to all of the information contained in

the approved Minimum Equipment List through printed or other means approved by the Director in the certificate holder operations specifications. An approved Minimum Equipment List, as authorized by the operations specifications, constitutes an approved change to the type design without requiring re-certification.

(3) The approved Minimum Equipment List must:

(i) Be prepared in accordance with the limitations specified in Paragraph (b) of this section.

(ii) Provide for the operation of the airplane with certain instruments and equipment in an inoperable condition.

(4) Records identifying the inoperable instruments and equipment and the information required by Paragraph (a)(3)(ii) of this section must be available to the pilot.

(5) The airplane is operated under all applicable conditions and limitations contained in the Minimum Equipment List and the operations specifications authorizing use of the Minimum Equipment List.

(b) The following instruments and equipment may not be included in the Minimum Equipment List:

(1) Instruments and equipment that are either specifically or otherwise required by the airworthiness requirements under which the airplane is type certificated and which are essential for safe operations under all operating conditions.

(2) Instruments and equipment required by an airworthiness directive to be in operable condition unless the airworthiness directive provides otherwise.

(3) Instruments and equipment required for specific operations by this part.

(c) Notwithstanding Paragraphs (b)(1) and (b)(3) of this section, an airplane with inoperable instruments or equipment may be operated under a special flight permit under Sections 21.197 and 21.199 of the CASRs.

Air Traffic Services Provider

The *Perusahaan Umum Lembaga Penyelenggara Pelayanan Navigasi Penerbangan Indonesia* (AirNav Indonesia) is the Air Traffic Services (ATS) provider within Indonesia. The ATS in Jakarta is provided by AirNav Indonesia branch office Jakarta Air Traffic Service Center (JATSC) which held a valid Air Traffic Services provider certificate. The services provided were aerodrome control service, approach control service, area control service, aeronautical communication service, and flight information services.

The approach control service for LNI610 flight was provided by the Terminal East and Arrival controllers utilizing surveillance control (radar service). The Arrival controller for normal flight operation provides approach control service for arriving aircraft to Soekarno-Hatta International Airport.

Procedures Related to Emergency Condition

According to the JATSC Standard Operation Procedure for Approach Control Services chapter 6.2 mentioned that condition when flight crew report any instrument malfunction might be suspected or classified as an emergency situation. The chapter 6.2.3.2 described procedure shall be followed to handle emergency situation as follows:

a. Immediately and briefly acknowledge the emergency situation.

b. Immediately report the situation to the supervisor.

c. If flight crew declare emergency, controller should take necessary action to ascertain: • Aircraft identification and type

• The type of emergency

• Aircraft position and altitude

• Type of assistance needed

• Enlist the aid of any other ATS unit which may be able to provide assistance to the aircraft

• Provide the flight crew with any information requested, such as weather information minimum safe altitudes and details on suitable aerodromes to land

d. Determine the appropriate assistance based on the available information.

e. Any condition indicates that the flight crew experiencing emergency condition but hesitate to take decision, the controller shall consider the worst possibilities or risk prior to issue instruction or suggestion.

f. Obtain from the operator or the flight crew of information that may be relevant such as: number of persons on board, amount of fuel remaining, possible presence of hazardous materials and the nature thereof.

g. Notify the appropriate ATS units and authorities.

h. Record all received information and action had been done.

i. Inform all aircraft who operate near the emergency aircraft.

j. Instruct all other aircraft to fly near the location of emergency aircraft and relay controller instruction if the emergency aircraft is unable to receive the instruction and to monitor the Emergency Locator Beacon (ELBA).

k. Changes of radio frequency and SSR code should be avoided if possible and should normally be made only when or if an improved service can be provided to the aircraft concerned.

l. Maneuvering instructions to an aircraft experiencing engine failure should be limited to a minimum.

m. Instruction to aircraft should be made briefly and clear.

n. If it is required, request assistance to other controllers or supervisor.

CHAPTER 5

FEDERAL AVIATION ADMINISTRATION (FAA) AIRCRAFT CERTIFICATION

Directorate General of Civil Aviation in Indonesia

The Directorate General Civil Aviation (DGCA) is part of Ministry of Transportation responsible for administer the civil aviation in Indonesia. The duties and responsibility of the DGCA is stated in the Transport Minister Decree 122/2018.

One of the directorates within the DGCA is Directorate of Airworthiness and Aircraft Operations (DAAO) which has sub-directorates of Standards, Aircraft Certification, Licensing, Airworthiness and Operation.

The Type Certificate Validation (TC Validation) is part of the sub directorate Aircraft Certification responsibilities whenever a type of aircraft will be operated by Indonesia aircraft operator.

The DAAO also approve the operator eligibility in operation and maintenance as described in the Authorizations, Conditions and Limitations (ACL) document including the Continuous Airworthiness Maintenance Program (CAMP).

The oversight to the aircraft operator or maintenance organization is performed by sub-directorate Airworthiness and Operation which is conducted by the annual audit and periodic surveillance. The DAAO assigned Principle Airworthiness Inspector (PAI) and Principle Operation Inspector (POI) for each aircraft operator while the maintenance organization oversight only by the PAI.

During annual audit or surveillance, the PAI or POI will ensure that all procedures are conform to the requirement including documents revision.

Type Certification Process and Overview

The American Federal Aviation Administration (FAA) is responsible for prescribing minimum standards required in the interest of safety for the design, material, construction, quality of work, and performance of aircraft, aircraft engines, and propellers (Ref. 49USC44701). Product certification 21 is a regulatory process administered by the FAA to ensure that an aircraft manufacturer's product conforms with Federal Aviation Regulations (FAR). Successful completion of the certification process enables the FAA to issue a Type Certificate (TC) or an Amended Type Certificate. To obtain a TC or an Amended Type Certificate, the manufacturer must demonstrate to the FAA that the aircraft or product being submitted for approval complies with all applicable regulations. The FAA determines whether or not the applicant has met its responsibility to show compliance to the applicable regulations.

The Federal regulations that apply to type certification of transport-category aircrafts are 14 CFR Part 21, 25, 26, 33, 34, and 36. The Part 25 regulations are those concerned with the airworthiness standards for transport-category aircrafts and are organized into subparts A through G. These regulations represent the minimum standards for airworthiness; an applicant's design may exceed these standards and the applicant's tests and analyses may be more extensive than required by regulation. The specific applicable regulatory requirements and how compliance will be demonstrated is documented in an FAA approved certification plan.
During the certification of Boeing 737-8 MAX, there were multiple civil aviation authority certification who participated in "validation" of the design.

Certification Guidance

FAA Order 8110.4C, titled "Type Certification", prescribes the responsibilities and procedures the FAA must follow to certify new civil aircraft, aircraft engines, and propellers, or changes thereto, as required by 14 of the CFR Part 21. This order is primarily written for internal use by the FAA, its designees, and delegated organizations. The order provides procedures and policy for the type certification of products and, unless stated otherwise, the type certification process in this order applies to all U.S. TCs, including amended TCs and Supplement Type Certificate (STCs).

Type Certification Process

FAA Order 8110.4C contains a section that presents a high-level flow diagram of the certification events that typically make up the life cycle an aircraft. The diagram is meant to explain the type certification process, not to dictate precisely how the project should flow. Although the model shows the proper sequence of events for certificating a product, the various aspects of the project generally progress through the process at different times and at different rates. The model divides the product's type certification life cycle into phases based on The FAA and Industry Guide to Product Certification. For each of the certification events identified on the flow diagram, the Order also provides information describing each event identifies expectations and develops specific interface procedures between the applicant and the FAA.

During a meeting with the NTSB24, the FAA provided a high-level overview of the certification process for an amended type design program. The briefing indicated that the applicant would start by conducting familiarization briefings and submitting the following to the FAA: a Certification Project Notification (CPN), a Program Notification Letter (PNL) and a Master Certification Plan (MCP). These documents detail the changes and identify the regulatory requirements and policies that are applicable; they also identify areas of change associated with the FAA airworthiness directives.

FAA Certification Office

The FAA has 10 Aircraft Certification Offices (ACO) which are responsible for approving the design certification of aircraft, aircraft engines, propellers, and replacement parts for those products. There are also specialized certification offices which include the Engine Certification Office (ECO), the Military Certification Office (MCO), the Boeing Aviation Safety Oversight Office (BASOO), and the Delegation Systems Certification Office (DSCO). FAA's BASOO responsibilities include oversight of Boeing's Organization Designation Authorization (ODA), involvement in certification of safety critical areas as well as novel and unusual designs and assisting foreign Civil Aviation Authorities (CAAs) in validation of Boeing products. The BASOO was responsible for the certification oversight and approval for the Boeing 737-8 (MAX).

Certification Basis for Changed Aviation Products

The certification basis for changed aeronautical products allows an aircraft manufacturer to introduce a derivative model as a design update on a previously certificated aircraft and add the changed product onto an existing TC. The FAA approves such changes if it finds that the changes are not significant enough to warrant application for a new TC. This process enables a manufacturer to introduce derivative aircraft models without having to resubmit the entire aircraft design for certification review. The manufacturer can use the results of some of the analyses and testing from the original type certification to demonstrate compliance, in which case the regulations that were in effect on the date of the original TC apply.

Title 14 CFR 21.101, Subpart D, specifies the requirements for demonstrating airworthiness compliance for changed aeronautical products. The current revision of 14 CFR 21.101, amendment 21.92, which became effective on April 16, 2011, states that an application for a changed aeronautical product to be added to a TC "must show that the changed product complies with the airworthiness requirements applicable to the category of the product in effect on the date of the application." This regulation is more specific than previous revisions regarding what can be used from the original certification basis in an application for a derivative model involving a major change.

On April 25, 2003, the FAA issued FAA Order 8110.48, How to Establish the Certification Basis for Changed Aeronautical Products, which provides the procedures that the FAA utilize for determining the certification basis for changes to type certificated products including changes made through an amended Type Certificate which is the method utilized for the G-IV. The handbook refers to FAA Advisory Circular 21.101-1, establishing the Certification Basis of Changed Aeronautical Products, which contains an acceptable means, but not the only means, to comply with 14 CFR 21.101. On July 21, 2107, this Order 8110.48 was cancelled and replaced by Order 8110.48A.

System Safety Assessment Process

As part of the process for developing and certifying a safety-critical system, individual system safety assessments are conducted to assure the system designs meet their safety requirements and support the aircraft level safety assessment. Safety assessments are conducted by the applicant, and its suppliers, and are reviewed and accepted by the FAA. The safety assessment process is outlined in AC 25.1309-1A and described in detail in

the SAE Aerospace Recommended Practice (ARP) 4761. SAE ARP 4761 describes the formal procedure as follows:
• Functional Hazard Assessment (FHA) to address hazard identification and preliminary risk analysis

• Preliminary System Safety Assessment (PSSA) to analysis the contribution and interaction of the subsystem to system hazard

• System Safety Assessment (SSA) to assess the results of design and implementation, ensuring that all safety requirements are met.

Functional Hazard Assessment (FHA):

A functional hazard assessment (FHA) is a systematic examination of a system's functions and purpose, and it typically provides the initial, top-level assessment of a design and addresses the operational vulnerabilities of the system function. The FHA is therefore used to establish the safety requirements that guide system architecture design decisions. Performed independently of any specific design, an FHA evaluates what would occur if the function under question was lost or malfunctioned and classifies that effect to prioritize focus on the most serious outcomes. An FHA is conducted early in the design and development cycle to identify failure conditions and classify them by severity, beginning at the aircraft level and working down to individual systems.

FAA Advisory Circular AC 25.1309-1A, dated June 21, 1988 and SAE ARP4761 define the severity classes that are used to classify the effect of loss or malfunction as part of an FHA. AC 25.1309-1A defines the following three severity classes: catastrophic, major and minor, with the respective likelihoods, of extremely improbable (one-in-a billion/10^{-9} or less), improbable (one-in-ten million/10^{-7} or less), or no worse than probable (one-in-hundred thousand/10^{-5}). The differences among the classes are associated with effects on the aircraft, occupants, and crew. According to SAE ARP4761, the determination of the classification is accomplished by analyzing accident/incident data, reviewing regulatory guidance material, using previous design experience, and consulting with flight crews, if applicable. The failure condition severity classifications are provided in a table contained within this document and are defined as follows:

• Catastrophic:
  All failure conditions which prevent continued safe flight and landing.
• Severe-Major/Hazardous:

Large reductions in safety margins or functional capabilities. Higher workload or physical distress such that the crew could not be relied upon to perform tasks accurately or completely.

- Major:
Significant reduction in safety margins or functional capabilities. Significant increase in crew workload or in conditions impairing crew efficiency.

- Minor:
A slight reduction in safety margins. A slight increase in crew workload.

System Safety Assessments:

Safety assessments are a primary means of compliance for systems (as opposed to identifying structures or aircraft performance characteristics) that are critical to safe flight and operation. Safety assessments proceed in a stepwise, data-driven fashion, analogous to the system development process described above. Starting with aircraft functions, functional hazard assessments are performed to identify the failure conditions associated with each function. Systems functional hazard analyses are performed for system level functions. Preliminary safety assessments are performed as the system is developed adding more specific design and implementation detail to address specific hazards. The bottom-up verification by safety analysis starts with an analysis of the components of a system to ensure single failures do not result in significant effects. Combinations of failures are logically combined to develop probability of a failure and checked to ensure they are commensurate with the criticality of the failure condition. Thus, the final definition and characterization of a safety-critical system is verified by the result of the analyses conducted during a safety assessment.

Safety assessments are conducted by the applicant, and its suppliers, and are reviewed and accepted by the FAA. The safety assessment process is outlined in AC 25.1309-1A and described in detail in SAE ARP4761. Although the safety assessment process outlined in the AC is not mandatory, applicants who choose not to conduct safety assessments must demonstrate compliance in another, FAA-approved way (for example, by conducting ground or flight tests).

Organization Designation Authorization

In title 14, Code of Federal Regulations (CFR) United States of America Part 183, the Federal Aviation Administration (FAA) may delegate the specified functions to an organization on behalf of the Administrator related to engineering, manufacturing, operations, airworthiness, or maintenance.

In the Part 183 subpart D, the organization granted by the FAA for such delegation is referred as Organization Designation Authorization (ODA) which means the organization is authorized to perform certification functions on behalf of the FAA. FAA granted the Boeing Commercial Airplane (BCA) ODA in 2009. The delegated functions for a Type Certification ODA are:

- establishing and determining conformity of parts, assemblies, installations, test setups, and products (aircraft);
- finding compliance with airworthiness standards for new design, or major changes to design;
- ssuing special flight permits for operation of aircraft;
- issuing issues airworthiness approvals for articles (Export), and aircraft (Standard or Export).

Oversight and Delegation

Inspector General Audit Report:

According to a 2011 Office of Inspector General audit report[15], *"the Federal Aviation Administration (FAA) is responsible for overseeing numerous aviation activities designed to ensure the safety of the flying public. Recognizing that it is not possible for FAA employees to personally oversee every facet of aviation, public law allows FAA to delegate certain functions, such as approving new aircraft designs, to private individuals or organizations (approved by the FAA). Designees perform a substantial amount of critical work on FAA's behalf—for example, at one aircraft manufacturer, they made about 90 percent of the regulatory compliance determinations for a new aircraft design. FAA created the Organization Designation Authorization (ODA) program in 2005 to standardize its oversight of organizational designees"*.

According to FAA Order 8100.15A, 49 CFR 44702(d) allows the FAA to delegate to a qualified private person a matter related to issuing certificates, or related to the examination, testing, and inspection necessary to issue a certificate on behalf of the FAA Administrator as authorized by statute to issue under 49 CFR 44702(a).

Boeing applied for and was granted ODA. Boeing's ODA is authorized to select and appoint individuals to perform some of the delegated functions as representatives of FAA. The delegated functions for a Type Certification (TC) ODA are:

- establishing and determining conformity of parts, assemblies, installations, test setups, and products (aircraft);

- finding compliance with airworthiness standards for new design, or major changes to design;

- issuing special flight permits for operation of aircraft;

- issuing issues airworthiness approvals for articles (Export), and aircraft (Standard or Export)

Guidance for Delegation of Compliance Findings:

FAA Order 8110.4C, section 2.5, titled "Compliance Planning," discusses the FAA's involvement in a certification project, including providing guidance on oversight and delegation. According to the order, *"For planning purposes, the FAA's and the applicant's certification teams need to know in which aspects of the project the FAA intends involvement and at what level. The heavy workloads for FAA personnel limit involvement in certification activities to a small fraction of the whole. FAA type certification team members must review the applicant's design descriptions and project plans, determine where their attention will derive the most benefit, and coordinate their intentions with the applicant"*.

Paragraph (a)(1) of section 2.5 provides guidance to the FAA and applicant on the identification of critical safety items requiring direct FAA involvement in the findings of compliance. According to the paragraph, *"When a particular decision or event is critical to the safety of the product or to the determination of compliance, the FAA must be directly involved (as opposed to indirect FAA involvement by, for example, DERs10). Project team members must build on their experience to identify critical issues. Some key issues that will always require direct FAA involvement include rulemaking (such as for special conditions), development of issue papers, and compliance findings considered unusual or typically reserved for the FAA. While these items establish the minimum direct FAA involvement, additional critical safety findings must also be identified based on the safety impact or the complexity of the requirement or the method of compliance. Additional factors to consider in determining the areas of direct FAA involvement include the FAA's confidence in the applicant, the applicant's experience, the applicant's internal processes, and confidence in the designees"*.

Certification Basis:

According to an FAA Issue Paper[16], on January 27, 2012, the Boeing Company, located in Seattle, Washington, submitted an application to the BASOO to amend the Model 737 Type Certificate No. A16WE to include the new model Boeing 737-8 (MAX). The Boeing Company submitted their request via a letter dated April 8, 2016 to extend the application date to February 27, 2012. Due to late certification of the engines, the applicant submitted a second request via a letter dated December 14, 2016, to extend the application date to June 30, 2012. The Issue Paper indicated that these changes do not affect the previously agreed upon certification basis. Therefore, in accordance with Title 14, Code of Federal Regulations (14 CFR) part 21 requirements, the certification basis for the Model Boeing 737-8 (MAX) was established within the Issue Paper. Boeing's application for an amendment to the type certificate A16WE of the Model Boeing 737-8 (MAX) was effective for 5 years from the new date of application.

According to Type Certificate Data Sheet (TCDS)[17] A16WE, revision 64, dated October 10, 2018, the Boeing Company applied for a transport category amended type certificate for the Boeing 737-8 (MAX) aircraft On June 30, 2012 and type certificate approval for the Boeing 737-8 (MAX) aircraft was granted on March 8, 2017, under 14 CFR Part 25 (the airworthiness standards for transport-category airplanes). The Boeing 737-8 MAX aircraft was added as the most recent model in a series of derivative models (or "changed aeronautical products") that were approved and added to the Boeing type certificate (TC), originally issued for the Boeing 737-100 on December 15, 1967.

The applicable certification basis for the Boeing 737-8 (MAX) aircraft is Title 14, Code of Federal Regulations (14 CFR) part 25 as amended by Amendments 25-0 through 25-137, plus amendment 25-141 with exceptions permitted by 14 CFR 21.101.

The Boeing Model Boeing 737-8 (MAX) and Boeing 737-9 was granted an exception per 14 CFR 21.101(b) for § 25.795(c)(2) based on the demonstration and justification that security features were present in the type design. These security features must be in consideration in any subsequent type design change, modification, or repair to ensure the level of safety designed into the Boeing 737-8 (MAX) and 737-9 is maintained. In lieu of the following, compliance to § 25.795(c)(2), Amendment 25-127, may be shown:

Amended Type Certification (ATC) Application		January 2012
General Familiarization Meeting	(completed)	March 2012
Technical Familiarization Meetings	(completed)	May 2012

FAA Acceptance of Master Certification Basis Established	Certification Plan (G-1 Issue Paper)	November 2013 February 2014
FAA Acceptance of (related) Type Inspection Authorization	Detailed Certification Plans Approved	November 2016 August 2016
FAA Certification Flight Tests ATC Issuance	Complete	February 2017 March 2017

Maneuvering Characteristics Augmentation System (MCAS) Assessment

MCAS Assessment General:

As part of the MCAS development phase, in late 2012, Boeing performed a preliminary functional hazard assessment[18] of MCAS using piloted simulations in their full motion Engineering Flight Simulator (E-Cab). Several hazards were assessed at that time, however, this section of the report will focus only on the following two hazards: uncommanded MCAS operation up to its maximum authority (0.6 degrees of aircraft nose down stabilizer) and uncommanded MCAS operation equivalent to a 3 second stabilizer trim runaway[19].

Boeing used two scenarios to assess this hazard: a runaway at MCAS activation during a wind-up turn maneuver (operational envelope) and a wings-level recovery from a stabilizer runaway during level flight (normal flight envelope).

To perform these simulator tests, Boeing induced a stabilizer trim input that would simulate the stabilizer moving at a rate and duration consistent with the MCAS function. Using this method to induce the hazard resulted in the following: motion of the stabilizer trim wheel, increased column forces, and indication that the aircraft was moving nose down. Boeing indicated that this evaluation was focused on the flight crew response to uncommanded MCAS operation, regardless of underlying cause. Thus, the specific failure modes that could lead to uncommanded MCAS activation, such as an erroneous high AOA input to the MCAS, were not simulated as part of these functional hazard assessment validation tests. As a result, additional flight deck effects (such as IAS DISAGREE and ALT DISAGREE alerts and stick shaker activation) resulting from the same underlying failure (for example, erroneous AOA) were not simulated and were not documented in the stabilizer trim and autoflight safety assessment reports reviewed after the accident.

Original results of preliminary hazard assessment

Hazard	Hazard classification
Uncommanded MCAS operation up to its maximum authority	Major
Uncommanded MCAS function operation equivalent to 3 second mistrim	Major

The FHA evaluations were conducted by Boeing in their Engineering Cab using FAA guidance regarding flight crew response to flight control failures requiring trim input that is contained in FAA Advisory Circular AC25.7C[20]. In particular, Boeing uses the following assumptions in its flight controls FHAs:

• Uncommanded system inputs are readily recognizable and can be counteracted by overriding the failure by movement of the flight controls in the normal sense by the flight crew and do not require specific procedures.

• Action to counter the failure shall not require exceptional piloting skill or strength.

• The flight crew will take immediate action to reduce or eliminate increased control forces by re-trimming or changing configuration or flight conditions.

• Trained flight crew memory procedures shall be followed to address and eliminate or mitigate the failure.

Boeing advised that these assumptions are used across all Boeing models when performing functional hazard assessments of flight control systems and that these assumptions are consistent with the requirements contained in 14 CFR 25.671 & 25.672 and within the guidance contained in FAA Advisory Circular (AC) 25-7C for compliance evaluation of 14 CFR 25.143[21].

In March 2016, Boeing determined that MCAS should be revised to improve wings-level, flaps up, low Mach stall characteristics and identification. The MCAS was revised such that depending on AOA, it would be capable of commanding incremental stabilizer to a maximum of 2.5 degrees at low Mach decreasing to a maximum of 0.65 degrees at high Mach.

The requirements document also indicated that the preliminary functional hazard assessments of MCAS were re-evaluated by flight crew assessments in the motion simulator and by engineering analysis and determined to have not changed in hazard classification as a result of the increase in MCAS authority to 2.5 degrees.

for revised MCAS authority

Hazard	Hazard classification
Uncommanded MCAS function operation up to its maximum authority	Major*
Uncommanded MCAS function operation equivalent to 3 second mistrim **	Major

* Major Classification:

The uncommanded MCAS command to the maximum nose down authority at low Mach numbers was evaluated in the Boeing 737-8 (MAX) cab and rated as Minor. The high Mach uncommanded MCAS command and subsequent recovery is the critical flight phase in establishing the hazard rating for erroneous MCAS commands. According to Boeing, engineering analysis determined that the existing high Mach evaluations remain valid as the aerodynamic configuration had not changed significantly since the pre-flight evaluations, and the MCAS authority limit at high Mach did not change significantly in the flight test update. As the ratings for these high Mach evaluations were more severe than for low Mach, the overall flight envelope hazard ratings remain the same as the pre-flight evaluations.

** Piloted Simulation not Required:

According to Boeing, Engineering analysis determined no low Mach piloted simulation to be required as this failure is less critical than MCAS function operation to maximum authority. Stabilizer motion for three seconds would not reach maximum authority in low Mach conditions. The existing high Mach evaluations remain valid as the aerodynamic configuration has not changed significantly since the preflight evaluations, and the 3 second stabilizer motion is the same magnitude.

When assessing unintended MCAS activation in the simulator for the FHAs, the function was allowed to perform to its authority and beyond before flight crew action was taken to recover. Failures were able to be countered by using elevator alone. Stabilizer trim was available to offload column forces, and stabilizer cutouts were available but not required to counter failures. This was true both for the preliminary FHAs performed in 2012 and for the reassessment of the FHAs in 2016.

In a 2019 presentation to the investigation team, Boeing indicated that the MCAS hazard classification of "major" for uncommanded MCAS function (including up to the new authority limits) in the Normal flight envelope were based on the following conclusions:

- Unintended stabilizer trim inputs are readily recognized by movement of the stabilizer trim wheel, flight path change or increased column forces.

- Aircraft can be returned to steady level flight using available column (elevator) alone or stabilizer trim.

- Continuous unintended nose down stabilizer trim inputs would be recognized as a Stab Trim or Stab Runaway failure and procedure for Stab Runaway would be followed.

Indonesia Type Certificate Validation Process

Referring to Indonesia Aviation Act 1/2009 article 16, every aircraft, aircraft engine, and aircraft propeller designed and produced overseas and imported into Indonesia must obtain a type certificate validation. The Civil Aviation Safety Regulation (CASR) Part 21 regulates the compliance procedure for the Indonesia Aviation Act 1/2009 article 16 and it is outlined in the Directorate General of Civil Aviation (DGCA) Staff Instruction SI 21-03 Validation Procedures of Foreign Type Certificate (Aircraft, Engine and Propeller).

The Indonesia DGCA conducted the Type Certificate Validation (TC Validation) for the Boeing 737-8 (MAX) on 3 – 12 April 2017 in Boeing facility at Renton Seattle Washington DC. At the same time three foreign civil aviation authorities (Malaysia, Japan and United Arab Emirates) also conducted the similar TC Validation for the Boeing 737-8 (MAX).

The Indonesia TC Validation team consisted of Team Leader, Powerplant System inspector, Avionic System inspector and Flight Test inspector.
The agenda of the TC Validation were system familiarization presented by Boeing and certification discussion based on the FAA issue paper G-1 (Federal Aviation Regulation Part 25 Certification Basis), paper G-2 (compliance checklist) and paper G-3 (environmental considerations). The Maneuvering Characteristic Augmentation System (MCAS) was discussed during the system familiarization Boeing 737-8 (MAX).

Since the time frame was not sufficient the Indonesia DGCA requested time extension for discussion of additional requirements of CASR. The discussion resulted in nine additional requirements identified as follows:
- PM01 (Certification Basis of CASR Part 25),
- PM02 (Test Documents),

- CS01-737 (Bilingual Placard),
- CS02-8-Max (Least Risk Bomb Location),
- CS03-737 (Ditching Requirement as per ICAO Annex 8),
- AV-01-Max (Electrical System),
- AV-02-Max (APU door system test),
- PP-01-Max (Early ETOPS) and
- PM-04-016-17A (Type Certification Data Sheet – TCDS).

The TC Validation was finalized in May 2017 and the DGCA issued the TC Validation document referred as A068. However, if there was system which had not been verified during the TC Validation process while the aircraft had been in service, the DGCA might issue the exemption or update the TC Validation document.

Airworthiness Regulations

Airworthiness Standard for Type Certificate

Aircraft type certificate is issued by civil aviation authority to ensure the aircraft is manufactured in accordance with approved design and complies with a product meets its type design and is in a condition for safe operation. The airworthiness standards for the issue of type certificates, and changes to those certificates, for transport category airplanes in Indonesia is described in the CASR Part 25 Airworthiness Standards: Transport Category and in the United States of America described in Federal Aviation Regulation (FAR) Part 25. Both CASR Part 25 and 14 FAR Part 25 contained same standards on same subparts number.
The relevant subparts were as follows:

Applicability
(a) This part prescribes airworthiness standards for the issue of type certificates, and changes to those certificates, for transport category airplanes.
(b) Each organization who applies under CASR Part 21 for such a certificate or change must show compliance with the applicable requirements in this part.

General.
(a) The airplane must be safely controllable and maneuverable during—
(1) Takeoff;
(2) Climb;
(3) Level flight;
(4) Descent; and

(5) *Landing.*

(b) It must be possible to make a smooth transition from one flight condition to any other flight condition without exceptional piloting skill, alertness, or strength, and without danger of exceeding the airplane limit-load factor under any probable operating conditions, including—

(1) The sudden failure of the critical engine;
(2) For airplanes with three or more engines, the sudden failure of the second critical engine when the airplane is in the en route, approach, or landing configuration and is trimmed with the critical engine inoperative; and
(3) Configuration changes, including deployment or retraction of deceleration devices.
(c) The airplane must be shown to be safely controllable and maneuverable with the critical ice accretion appropriate to the phase of flight defined in appendix C, and with the critical engine inoperative and its propeller (if applicable) in the minimum drag position:

(1) At the minimum V2 for takeoff;
(2) During an approach and go-around; and
(3) During an approach and landing.
(d) The following table prescribes, for conventional wheel type controls, the maximum control forces permitted during the testing required by paragraphs (a) and (c) of this section:

Force, in pounds, applied to the control wheel or rudder pedals	Pitch	Roll	Yaw
For short term application for pitch and roll control — two hands available for control	75	50	-
For short term application for pitch and roll control — one hand available for control	50	25	-
For short term application for yaw control	-	-	150
For long term application	10	5	20

(e) Approved operating procedures or conventional operating practices must be followed when demonstrating compliance with the control force limitations for short term application that are prescribed in paragraph (d) of this section. The airplane must be in trim, or as near to being in trim as practical, in the preceding steady flight condition. For the takeoff condition, the airplane must be trimmed according to the approved operating procedures.

Out-of-trim characteristics.
  (a) From an initial condition with the airplane trimmed at cruise speeds up to VMO/MMO, the air-plane must have satisfactory maneuvering stability and controllability with the degree of out-of-trim in both the airplane nose-up and nose-down directions, which results from the greater of —

(1) A three-second movement of the longitudinal trim system at its normal rate for the particular flight condition with no aerodynamic load (or an equivalent degree of trim for airplanes that do not have a power-operated trim system), except as limited by stops in the trim system, including those required by §25.655(b) for adjustable stabilizers; or
(2) The maximum mistrim that can be sustained by the autopilot while maintaining level flight in the high speed cruising condition.
(b) In the out-of-trim condition specified in paragraph (a) of this section, when the normal acceleration is varied from +1 g to the positive and negative values specified in paragraph (c) of this section —

(1) The stick force vs. g curve must have a positive slope at any speed up to and including VFC/MFC; and
(2) At speeds between VFC/MFC and VDF/MDF the direction of the primary longitudinal control force may not reverse.

  (b) Except as provided in paragraphs (d) and (e) of this section, compliance with the provisions of paragraph (a) of this section must be demonstrated in flight over the acceleration range —
  (c)
(1) -1 g to +2.5 g; or
(2) 0 g to 2.0 g, and extrapolating by an accept-able method to -1 g and +2.5 g.

(d) If the procedure set forth in paragraph (c)(2)of this section is used to demonstrate compliance and marginal conditions exist during flight test with regard to reversal of primary longitudinal control force, flight tests must be accomplished from the normal acceleration at which a marginal condition is found to exist to the applicable limit specified in paragraph (b)(1) of this section.

(e) During flight tests required by paragraph (a)of this section, the limit maneuvering load factors prescribed in §§25.333(b) and 25.337, and the maneuvering load factors associated with probable inadvertent excursions beyond the boundaries of the buffet onset envelopes determined under§25.251(e), need not be exceeded. In addition, the entry speeds for flight test demonstrations at normal acceleration values less than 1 g must be

limited to the extent necessary to accomplish a recovery without exceeding VDF/MDF. (f) In the out-of-trim condition specified in para-graph (a) of this section, it must be possible from an overspeed condition at VDF/MDF to produce at least 1.5

g for recovery by applying not more than 125 pounds of longitudinal control force using either the primary longitudinal control alone or the primary longitudinal control and the longitudinal trim system. If the longitudinal trim is used to assist in producing the required load factor, it must be shown at VDF/MDF that the longitudinal trim can be actuated in the airplane nose-up direction with the primary surface loaded to correspond to the least of the following airplane nose-up control forces:

(1) The maximum control forces expected in service as specified earlier
(2) The control force required to produce 1.5 g.
(3) The control force corresponding to buffeting or other phenomena of such intensity that it is a strong deterrent to further application of primary longitudinal control force.

25.1309 Equipment, Systems, and Installations

(a) The equipment, systems, and installations whose functioning is required by this Decree, must be designed to ensure that they perform their intended functions under any foreseeable operating condition.

(b) The airplane systems and associated components, considered separately and in relation to other systems, must be designed so that-

(1) The occurrence of any failure condition which would prevent the continued safe flight and landing of the airplane is extremely improbable, and

(2) The occurrence of any other failure conditions which would reduce the capability of the airplane or the ability of the crew to cope with adverse operating conditions is improbable.

(c) Warning information must be provided to alert the crew to unsafe system operating conditions, and to enable them to take appropriate corrective action. Systems, controls, and associated monitoring and warning means must be designed to minimize crew errors which could create additional hazards.

(d) Compliance with the requirements of paragraph (b) of this section must be shown by analysis, and where necessary, by appropriate ground, flight, or simulator tests. The analysis must consider-

(1) Possible modes of failure, including malfunctions and damage from external sources.
(2) The probability of multiple failures and undetected failures.
(3) The resulting effects on the airplane and occupants, considering the stage of flight and operating conditions, and
(4) The crew warning cues, corrective action required, and the capability of detecting faults.

(e) In showing compliance with paragraphs (a) and (b) of this section with regard to the electrical system and equipment design and installation, critical environmental conditions must be considered. For electrical generation, distribution, and utilization equipment required by or used in complying with this chapter, except equipment covered by Approved Technical Specification or Technical Standard Orders containing

environmental test procedures, the ability to provide continuous, safe service under foreseeable environmental conditions may be shown by environmental tests, design analysis, or reference to previous comparable service experience on other aircrafts
(f) EWIS must be assessed in accordance with the requirements of sec.25.1709.
25.1329 Flight guidance system.
(g) Under any condition of flight appropriate to its use, the flight guidance system may not produce hazardous loads on the airplane, nor create hazardous deviations in the flight path. This applies to both fault-free operation and in the event of a malfunction, and assumes that the pilot begins corrective action within a reasonable period of time.
25.1585 Operating Procedures
(a) Operating procedures must be furnished for—
(1) Normal procedures peculiar to the particular type or model encountered in connection with routine operations;
(2) Non-normal procedures for malfunction cases and failure conditions involving the use of special systems or the alternative use of regular systems; and
(3) Emergency procedures for foreseeable but unusual situations in which immediate and precise action by the crew may be expected to substantially reduce the risk of catastrophe.
(b) Information or procedures not directly related to airworthiness or not under the control of the crew must not be included, nor must any procedure that is accepted as basic airmanship.
(c) Information identifying each operating condition in which the fuel system independence prescribed in Sec. 25.953 is necessary for safety must be furnished, together with instructions for placing the fuel system in a configuration used to show compliance with that section.
(d) The buffet onset envelopes, determined under Sec. 25.251 must be furnished. The buffet onset envelopes presented may reflect the center of gravity at which the airplane is normally loaded during cruise if corrections for the effect of different center of gravity locations are furnished.

The Responsibility for Airworthiness

CASR Part 91 General Operating and Flight Rules are applicable for the operation of aircraft (other than moored balloons, kites, unmanned rockets, and unmanned free balloons, and ultra-light vehicles) within Indonesia territory.
The CASR Part 91 subpart 91.7 described:

  (a) No person may operate a civil aircraft unless it is in an airworthy condition.

  (b) The pilot in command of a civil aircraft is responsible for determining whether that aircraft is in condition for safe flight. The pilot in command shall discontinue the flight when un-airworthy mechanical, electrical, or structural conditions occur. (a) Each certificate holder is primarily responsible for—

The CASR Part 121 Certification and Operating Requirements: Domestic, Flag, and Supplemental Air Carriers, related to regulation for delegating aircraft maintenance to AMO is as follow:

CASR 121.363 Responsibility for Airworthiness
(1) The airworthiness of its aircraft, including airframe, aircraft engines, propellers, appliances, and parts thereof;

(2) The performance of the maintenance, preventive maintenance, and alteration of its aircraft, including airframes, aircraft engines, propellers, appliances, emergency equipment, and parts thereof, in accordance with its manual and the related regulations; and

(3) Obtaining and assessing the continuing airworthiness information and recommendations from the organizations responsible for the type design.

(b) A certificate holder may make arrangements with another person for the performance of any maintenance, preventive maintenance, or alterations. However, this does not relieve the certificate holder of the responsibility specified in paragraph (a) of this section.

Flight Crew Reporting Procedure

The ICAO Annex 6 subchapter 4.5.4 mentions that the pilot in command shall be responsible for reporting all known or suspected defects in the aeroplane to the operator at the termination of the flight.

According to the CASR Part 121 subchapter 121.563 Reporting Mechanical Irregularities, Captain shall ensure that all mechanical irregularities occurring during flight time are entered in the maintenance log of the aircraft at the end of that flight time. Before each flight the captain shall ascertain the status of each irregularity entered in the log at the end of the preceding flight.

Radiotelephony Procedure for Urgency or Distress Condition

The Manual of Standard CASR Part 170-02: Aeronautical Communication Procedures subchapter 5.3 described distress and urgency radiotelephony communication procedures.

5.3.1.1 Distress and urgency traffic shall comprise all radiotelephony messages relative to the distress and urgency conditions respectively. Distress and urgency conditions are defined as:

a. Distress: a condition of being threatened by serious and/or imminent danger and of requiring immediate assistance.

b. Urgency: a condition concerning the safety of an aircraft or other vehicle, or of some person on board or within sight, but which does not require immediate assistance.

5.3.1.2 The radiotelephony distress signal MAYDAY and the radiotelephony urgency signal PAN PAN shall be used at the commencement of the first distress and urgency communication respectively.

5.3.1.3 The originator of messages addressed to an aircraft in distress or urgency condition shall restrict to the minimum the number and volume and content of such messages as required by the condition.

5.3.1.5 Distress and urgency traffic shall normally be maintained on the frequency on which such traffic was initiated until it is considered that better assistance can be provided by transferring that traffic to another frequency.

5.3.1.6 In cases of distress and urgency communications, In general, the transmissions by radiotelephony shall be made slowly and distinctly, each word being clearly pronounced to facilitate transcription.

5.3.2.4.1 The distress communications have absolute priority over all other communications, and a station aware of them shall not transmit on the frequency concerned, unless:

a) the distress is cancelled or the distress traffic is terminated;
b) all distress traffic has been transferred to other frequencies;
c) the station controlling communications gives permission;
d) it has itself to render assistance.

a. *Safety policy and objectives*
b. *Safety Risk Management*
c. *Safety assurance*
d. *Safety promotion*

Safety Management System Standard for Aircraft Operator

ICAO Annex 19 subchapter 4.1 requires aircraft operator to establish Safety Management System (SMS) as systematic approach to manage safety. The approach is designed to continuously improve safety performance through: the identification of hazards, the collection and analysis of safety data and safety information, and the continuous assessment of safety risks. The ICAO Annex 19 Appendix 2 described the

minimum requirement for SMS framework which must comprise four components and 12 elements, as follows:
- Management commitment and responsibility
- Safety accountabilities
- Appointment of key safety personnel
- Coordination of emergency response planning
- SMS documentation
- Hazard22 identification
- Safety risk assessment and mitigation
- Safety performance monitoring and measurement
- The management of change
- Continuous improvement of the SMS
- Training and education
- Safety communication

Hazard Identification as Part of Safety Risk Management

Hazard identification is the first step of Safety Risk Management (SRM), the ICAO Annex 19 Appendix 2, described that aircraft operator must develop and maintain a process that ensures that hazards associated with flight operations are identified based on a combination of reactive, proactive and predictive methods of safety data collection.

The ICAO Document 9859 provided guidelines to develop SMS within organization including aircraft operator[22]. The subchapter 2.5.2.10 of the document describes two main methodologies for identifying hazards, as follows:

a. Reactive, which involves analysis of past outcomes or events. Hazards are identified through investigation of safety occurrences. Incidents and accidents

are an indication of system deficiencies and therefore can be used to determine which hazard(s) contributed to the event.

b. Proactive, which involves collecting safety data of lower consequence events or process performance and analyzing the safety information or frequency of occurrence to determine if a hazard could lead to an accident or incident. The safety information for proactive hazard identification primarily comes from flight data analysis (FDA) programs, safety reporting systems and the safety assurance function. a. organizational safety policies and safety objectives;

b. organizational roles and responsibilities related to safety;

c. basic SRM principles;

d. safety reporting systems;

e. the organization's SMS processes and procedures; and

f. human factors.

The following explanations regarding the hazard identification process are excerpted from the ICAO Document 9859 subchapter 9.4.4.

Safety reporting system is one of the main internal sources within aircraft operator to identify hazard, especially a voluntary safety reporting system. Personnel at all levels and across all disciplines are encouraged to identify and report hazards and other safety issues through their safety reporting systems.

Safety reporting systems should be readily accessible to all personnel. A paper-based, web-based or desktop form can be used depending on the situation. Having multiple entry methods available maximizes the likelihood of staff engagement. Everyone should be made aware of the benefits of safety reporting and what should be reported.

Training and Education as Part of Safety Promotion

According to ICAO Annex 19 Appendix 2, aircraft operator must develop and maintain safety training program which ensures personnel are trained and competent to perform their SMS duties. The ICAO Document 9859 subchapter 9.6.4 described personnel who are trained and competent to perform their SMS duties, regardless of their level in the organization, is an indication of management's commitment to an effective SMS. The training program should include initial and recurrent training requirements to maintain competencies. Initial safety training should consider, as a minimum, the following:

Indonesia Requirement for Safety Management System

The national requirement in Indonesia regarding the development of SMS is described in the Civil Aviation Safety Regulation (CASR) Part 19. The subpart 19.17 (b) requires aircraft operator to have in place a SMS which can identifies and assess safety hazards includes to mitigate its risk.

The subpart 19.19 described that aircraft operator must define organization safety policy which includes commitment to encourage employees to report safety issues.

In regards with the hazard identification, the subpart 19.31 requires aircraft operator to develop and maintain formal means for effectively collecting, recording, acting on and generating feedback about hazards in operations, which combine reactive, proactive and predictive methods of safety data collection.

Formal means of safety data collection shall include mandatory, voluntary and confidential reporting systems.

The voluntary reporting is described in subpart 19.59 which requires aircraft operator to establish system to facilitate the collection of occurrences that may not be captured by the mandatory reporting system and other safety-related information which is perceived by the reporter as an actual or potential hazard to aviation safety.

Serious Incident within Indonesia Territory

According to the Aviation Law Number 1 of 2009 and Government Decree Number 62 of 2013 described that KNKT have responsibility to conduct investigation on serious incident of civil aircraft occurred within the territory of Republic of Indonesia.

The CASR Part 830 subpart 830.2 defines serious incident as:

An incident involving circumstances indicating that there was a high probability of an accident and associated with the operation of an aircraft which, in the case of a manned aircraft, takes place between the time any person boards the aircraft with the intention of flight until such time as all such persons have disembarked, or in the case of an unmanned aircraft, takes place between the time the aircraft is ready to move with the purpose of flight until such time as it comes to rest at the end of the flight and the primary propulsion system is shut down.

The Appendix B of the CASR Part 830 described that system failures, weather phenomena, operations outside the approved flight envelope or other occurrences which caused or could have caused difficulties controlling the aircraft is included in the list examples of serious incident.

Once serious incident of civil aircraft occurred within Indonesia territory, the CASR Part 830 subpart 830.06 requires person, organization or enterprise engaged in or offering to engage in an aircraft operation, with minimum delay and by the most suitable and quickest means available, must report to the *Komite Nasional Keselamatan Transportasi* (KNKT).

CHAPTER 6

ADDITIONAL INFORMATION

PK-LQP Previous Flight

On 28 October 2018, the accident aircraft was operated as a passenger flight from I Gusti Ngurah Rai International Airport (WADD), Denpasar[23] to Jakarta as LNI043.

During preflight check, the Captain discussed with the engineer the rectification that had been performed on the aircraft. The engineer informed the Captain that the aircraft had SPD and ALT flags on the Captain's PFD and the left AOA sensor had been replaced and tested accordingly. The Captain was convinced by the explanation from the engineer and the statement on the Aircraft Flight Maintenance Log (AFML) that the problem had been resolved.

The Captain conducted the crew briefing and stated that he would act as Pilot Flying on the flight to Jakarta. During the briefing the Captain mentioned the replacement of the left AOA sensor. On this flight, a dead heading crew, first officer of Lion Air Group, rated with Boeing 737-8 (MAX) was seated in the cockpit jump seat.

The flight departed about 1420 UTC which was originally scheduled at 1130 UTC, and during takeoff the flight crew did not notice any abnormalities. The aircraft departed from Denpasar with two pilots, five flight attendants and 182 passengers.

The crew recalled that about two seconds after landing gear retraction, the Takeoff Configuration Warning[24] activated. However, this warning only occurs when the main gear is still on the ground.

About 400 feet, the Captain noticed on the Primary Flight Display (PFD)[25] that the IAS Disagree warning appeared and the stick shaker activated. The DFDR showed the stick shaker activated just after airborne. Following that indication, the Captain maintained a pitch of 15° and the existing takeoff thrust setting. The stick shaker remained active throughout the flight for about 96 minutes until landing.

The Captain cross checked the PFDs and determined his were erroneous. He handed over control to the FO (who had good instruments) and called for the airspeed unreliable memory items. After the transfer of control, the Captain cross checked the PFDs with the standby instrument and determined that the Captain's PFD had the incorrect information. The Captain then switched on the right flight director (F/D) so the FO would have a normal display.

After completion of the memory items, the Captain looked down to take the QRH, the dead heading pilot informed to the Captain that the aircraft was diving down. The Captain commanded the FO to pitch up and to follow the F/D command. The FO replied that the control column was heavy. The Captain advised the FO to re-trim the aircraft as required.

At 14:25:46 UTC, the Captain declared an urgency message by calling "PAN PAN" to the Denpasar Approach controller and described that the reason was instrument failure and requested to maintain runway heading. The Denpasar Approach controller acknowledged the message and approved the flight crew request. A few second later, the Denpasar Approach controller asked the flight crew whether the flight would return to Denpasar and the Captain responded "standby".

The Captain noticed that a few seconds after the FO had discontinued the electrical trim input, the stabilizer trim was automatically trimming the aircraft nose down (AND). After three automatic AND trim occurrences, the FO commented that the control column was too heavy to hold back. The stabilizer moved without flight crew input, which the Captain considered the automatic trim inputs as a runaway stabilizer. The Captain performed the memory items for Runaway Stabilizer NNC and positioned the STAB TRIM CUTOUT switches in the Cut-Out position. The DFDR recorded at 14:28:08 UTC the automatic trim and manual trim movement stopped.

A few minutes later, the Captain re-engaged the STAB TRIM CUTOUT switches to the NORMAL position, and almost immediately the automatic AND trimming re-occurred. The Captain then moved the STAB TRIM CUTOUT switches back to the Cut-Out position and continued with manual trim for the remainder of the flight.

The Captain reported that he performed three Non-Normal Checklists (NNCs) consisting of Airspeed Unreliable, Altitude DISAGREE, and Runaway Stabilizer. None of the NNCs performed contained the instruction "Plan to land at the nearest suitable airport". The Captain decided to continue the flight since none of the NNCs gave instructions to land at the nearest suitable airport and despite the degraded flight instrumentation, flying without autopilot and auto-throttle, and a continuous activation of stick shaker, he convinced himself that the aircraft was able to fly to the scheduled destination. The Captain did not inform the Lion Air ground station in Denpasar about the problems as he assumed that the aircraft would be able to continue the flight to Jakarta.

At 14:32:31 UTC, the LNI043 Captain advised to the Denpasar Approach controller that the problem had been resolved and requested to continue the flight at flight level 290 without Reduced Vertical Separation Minima (RVSM)[26]. The Denpasar Approach controller then instructed LNI043 to climb to altitude of flight level 280 and contact Makassar Area Control Center (ACC) for further air traffic control (ATC) services.

The Captain noticed on the Captain's PFD, that the minimum speed and maximum speed red and black (barber pole tape) merged and appeared continuous from top to bottom on the speed tape display. Because of that display, the Captain concentrated more on the FO's PFD and monitored the FO during the flight to Jakarta.

At 14:43:36 UTC, the Upper West Madura (UWM) controller of Makassar ACC instructed the LNI043 to climb to altitude of 38,000 feet.

At 14:48:27 UTC, the Captain declared urgency message to the UWM controller and requested to maintain flight level 280 due to instrument failure. The UWM controller acknowledged and approved the flight crew request.

At 14:54:07 UTC, the UWM controller instructed LNI043 to contact Upper West Semarang (UWS) controller for further ATC services.

At 14:55:28 UTC, the LNI043 Captain made an initial call to the UWS controller and advised that the aircraft was maintaining altitude of 28,000 feet. The UWS controller acknowledged the flight crew information and requested the detail of the instrument failure. The LNI043 Captain advised that altimeter and autopilot failed and requested the UWS controller to relay information to Jakarta controller to request an uninterrupted descent. The UWS controller acknowledged the LNI043 request.

The flight crew recalled that along the route, the weather was clear. During flight arrival preparation, the Captain received Jakarta weather information which indicated light rain with visibility of 5,000 meters.

During the flight, the Captain perceived threat such as difficulty in communication due to stick shaker noise, sense of panic and mental pressure.

During the flight, the FA-1 came to the cockpit and the Captain informed FA-1 that the fasten seat belt sign would be kept on all the way due to the previous disturbance and the rest of the flight was due to weather condition. The aircraft landed using runway 25L at 1556 UTC.

The DFDR recorded that after landing the electric stabilizer trim was active, which indicated that STAB TRIM CUTOUT switches were moved back to the NORMAL position.

After parking, the Captain informed the engineer verbally, about the aircraft problem and filed on the Aircraft Flight Maintenance Log (AFML) that IAS (Indicated Air Speed) and ALT (altitude) DISAGREE and FEEL DIFF PRESS (Feel Differential Pressure) light problems occurred during the flight. The Captain did not mention the activation of stick shaker to the engineer as he believed that the activation was an outcome from the mentioned problems. The Captain did not report that the STAB TRIM CUTOUT guarded switches were positioned to CUTOUT during flight and after landing returned to the NORMAL position.

The Captain also reported the flight condition through the electronic reporting system of the company using Air Safety Report (ASR) form. The event was reported as follows:

Airspeed unreliable and ALT Disagree shown after takeoff, STS* also running to the wrong direction, suspected because of speed difference, identified that CAPT instrument was unreliable and handover control to FO. Continue NNC of Airspeed Unreliable and ALT Disagree. Decide to continue flying to CGK at FL280, landed safely runway 25L.

Note: STS = Speed Trim System

The ASR was filed to the reporting system on 29 October 2019 LT in the early morning and the report acknowledged by the SS Department about 0830 LT.

The Captain also attempted to reach the Duty Management Pilot through the company group network (online messaging group application) to report the flight condition. The Duty Management Pilot responded; then the Captain sent the information written in the ASR, to the Safety and Security Department, as the Duty Management Pilot was in Jeddah, Saudi Arabia.

Similar Occurrence

On 10 March 2019, a Boeing 737-8 (MAX) registered ET-AVJ was being operated by Ethiopian Airlines for scheduled passenger flight from Addis Ababa Bole International Airport (HAAB), Ethiopia to Jomo Kenyatta International Airport (HKJK), Kenya with flight number ET-302.

Shortly after departure, the DFDR recorded the AOA sensor value became erroneous and the left stick shaker activated and remained active until near the end of the flight. The airspeed and altitude of the left air data system began to deviate from the corresponding right side values. The flight crew lost control of the aircraft and crashed at 28 Nm South East of Addis Ababa Bole International Airport Ethiopia.

The ICAO Annex 13 investigation is carried out by Ethiopia Aircraft Accident Investigation Bureau (AIB). The initial findings summarized from the preliminary report published by the Ethiopia AIB were as follows:

- The takeoff roll appeared normal, including normal values of left and right AOA.

- Shortly after airborne, the DFDR recorded left and right AOA value deviated and reached value of 74.5° while the right AOA value reached 15.3°. Afterwards, the stick shaker activated and remained active until near the end of the flight.

- After autopilot engagement, there were small amplitude roll oscillations accompanied by lateral acceleration, rudder oscillations and slight heading changes; these oscillations also continued after the autopilot disengaged.

- After the autopilot disengaged, the DFDR recorded automatic aircraft nose down (AND) trim command four times without flight crew's input. As a result, three motions of the stabilizer trim were recorded. The FDR data also indicated that the crew utilized the electric manual trim to counter the automatic AND input.

- The crew performed runaway stabilizer checklist and put the STAB TRIM cutout switch to cutout position and confirmed that the manual trim operation was not working.

Boeing Record on Stick Shaker Activation

The Boeing recorded 30 flights experienced stick-shaker on all Boeing model, in period of 2001 to 2018. The data shows 27 occurrences on Boeing 737 in which 21 flights returned, 5 flights continued, 3 flights diverted to the nearest airport and 1 flight where flight crew decision could not be determined. Three other occurrences were on Boeing 757 and Boeing 767 where all flights returned.

Among all flights experienced stick shaker recorded by Boeing, five flights were continued to destination. On the first flight experienced stick shaker, the initial left instrument problem occurred at about 6,000 feet and stick shaker occurred at 16,000 feet. The flight crew was sure that it was instrument problem and the weather along the route and destination were clear. The second flight experienced stick shaker during final approach and the flight crew elected to continue. The third flight stick shaker active after airborne and the flight crew performed Altitude Disagree and Unreliable Air Speed non-normal checklists. After completion the non-normal checklist, the stick shaker stopped and the flight crew elected to continue the flight. On the fourth flight, the stick shaker occurred after takeoff. The flight crew elected to continue as returning would result in overweight landing and the weather along the route was clear. About 40 minutes after takeoff, the flight crew pulled the circuit breaker of the affected control column with intention to eliminate noise and to make the stick shaker warning on the other side functioning normally. The fifth flight experienced stick shaker and the flight crew elected to continue was LNI043.

Situational Awareness

Situation Awareness in Aviation Systems - Mica R. Endsley (SA Technologies)

Situation awareness is formally defined as "the perception of the elements in the environment within a volume of time and space, the comprehension of their meaning and the projection of their status in the near future" (Endsley, 1988). Situation awareness therefore involves perceiving critical factors in the environment (Level 1 SA), understanding what those factors mean, particularly when integrated together in relation to the aircrew's goals (Level 2), and at the highest level, an understanding of what will happen with the system in the near future (Level 3). These higher levels of SA allow pilots to function in a timely and effective manner.

Level 1 SA — Perception of The Elements in The Environment

The first step in achieving SA is to perceive the status, attributes, and dynamics of relevant elements in the environment. A pilot needs to perceive important elements such as other aircraft, terrain, system status and warning lights along with their relevant characteristics. In the cockpit, just keeping up with all of the relevant system and flight data, other aircraft and navigational data can be quite taxing.

Level 2 SA — Comprehension of The Current Situation

Comprehension of the situation is based on a synthesis of disjointed Level 1 elements. Level 2 SA goes beyond simply being aware of the elements that are present, to include an understanding of the significance of those elements in light of one's goals. The aircrew puts together Level 1 data to form a holistic picture of the environment, including a comprehension of the significance of objects and events. For example, upon seeing warning lights indicating a problem during take-off, the pilot must quickly determine the seriousness of the problem in terms of the immediate air worthiness of the aircraft and combine this with knowledge on the amount of runway remaining in order to know whether it is an abort situation or not. A novice pilot may be capable of achieving the same Level 1 SA as more experienced pilots, but may fall far short of being able to integrate various data elements along with pertinent goals in order to comprehend the situation as well.

Level 3 SA — Projection of Future Status

It is the ability to project the future actions of the elements in the environment, at least in the very near term, that forms the third and highest level of situation awareness. This is achieved through knowledge of the status and dynamics of the elements and a comprehension of the situation (both Level 1 and Level 2 SA). Amalberti and Deblon (1992) found that a significant portion of experienced pilots' time was spent in anticipating possible future occurrences. This gives them the knowledge (and time) necessary to decide on the most favorable course of action to meet their objectives.

Crew Coordination

The research report by Gudela Grote & Enikö Zala-Mezö with title "The effects of different forms of coordination in coping with work load: Cockpit versus operating theatre" is summarized as follow

Coordination defined as tuning of interdependent work processes to promote concerted action towards a superordinate goal (Kieser & Kubicek, 1992) is needed for any

activity which cannot be carried out by one person and which cannot be subdivided into independent parts (Hackman & Morris, 1975).

Coordination is therefore a core activity in any work organization. As Tesluk et al. (1997, p. 197) formulate: "The essence of organizational action is the coordination, synchronization, and integration of the contributions of all organizational members into a series of collective responses." Studying accident and incident reports one can state that failures occur most often not because of technical or individual insufficiency but because a team fails to coordinate its mutual action (Hackman, 1993).

As a consequence of the degree and type of division of lab our and specialization more or less effort will be required for coordination and different kinds of coordination mechanisms will be more or less successful. Crucial in this respect is the type of interdependence created by the chosen division of labor in combination with the general demands of the task and the task environment. Generally, three types of interdependence of work activities are distinguished (e.g. Tesluk et al., 1997; Thompson, 1967; Van de Ven, Delbecq & Koenig, 1976):

Pooled interdependence is present, when system performance is an additive function of individual performance. The performance of other members of the system can have an effect on the work of the individual members but only indirectly through parallel contributions to a super ordinate goal.

Coordination in this case is usually achieved via centrally determined work programs which every individual has to follow independently and which assure that subtask serve the super ordinate goal.

Sequential interdependence is a unidirectional workflow arrangement, where individual performance depends on the proper fulfillment of prior subtasks. Synchronization is needed here based on centrally determined programs and plans that spell out the exact content and temporal requirements of subtask fulfillment.

Reciprocal interdependence means that information and results of work activities have to be exchanged between team members continuously. Coordination is mainly achieved via direct communication, be it in the form of personal directives or multilateral flow of communication between the individuals involved in self-regulated task performance.

FAA AC 120-71B

The FAA AC 120-71B provided guidance for the design, development, implementation, evaluation, and updating of standard operating procedures (SOP), and also guidance for pilot monitoring (PM) duties. The relevant part of the AC 120-71B was as follows:

CHAPTER 6. PILOT MONITORING

6.1 General. Several studies of crew performance, incidents, and accidents have identified inadequate monitoring and cross-checking as vulnerabilities for aviation safety. Effective monitoring and cross-checking can be the last barrier or line of defense against accidents because detecting an error or unsafe situation may break the chain of events leading to an accident. Conversely, when this layer of defense is absent, errors and unsafe situations may go undetected, potentially leading to adverse safety consequences. Flight crews must use monitoring to help them identify, prevent, and mitigate events that may impact safety margins. Therefore, it is imperative that operators establish operational policy and procedures on PM duties, including monitoring, and implement effective training for flight crews and instructors on the task of monitoring to help the PM expeditiously identify, prevent, and mitigate events that may impact safety margins.

This section describes effective monitoring, how to define and train PM duties, and integration of monitoring into SOPs. Additionally, the section discusses special considerations for monitoring auto-flight operations.

6.4 Defining Pilot Monitoring Duties. In a two-pilot operation, one pilot is designated as PF and one pilot is designated as PM. A review of operators' manuals indicates that the roles and associated tasks of the PF and PM are not always clearly defined. Each operator should explicitly define the roles of the PF and PM to include:

1. At any point in time during the flight, one pilot is the PF and one pilot is the PM.

2. The PF is responsible for managing, and the PM is responsible for monitoring the current and projected flightpath and energy of the aircraft at all times.

3. The PF is always engaged in flying the aircraft (even when the aircraft is under AP control) and avoids tasks or activities that distract from that engagement. If the PF needs to engage in activities that would distract from aircraft control, the PF should transfer aircraft control to the other pilot, and then assume the PM role.

4. Transfer of PF and PM roles should be done positively with verbal assignment and verbal acceptance to include a short brief of aircraft state.

5. The PM supports the PF at all times, staying abreast of aircraft state and ATC instructions and clearances.

6. The PM monitors the aircraft state and system status, calls out any perceived or potential deviations from the intended flightpath, and intervenes if necessary.

7. The PF provides a briefing to a pilot returning from a break. The briefing should include appropriate information to ensure the pilot returning from the break is updated on aircraft and systems states and current ATC instructions and assignments.

6.5 Operational Policies and Procedures. Operational policies and procedures should be reviewed or developed to ensure the division of duties and responsibilities between flight crew members protects the ability of the PF to control the flightpath. Assigning non-flightpath-related tasks to the PF should generally be avoided. Operational data should be collected and used to revise definitions of PF and PM roles and responsibilities to ensure their effectiveness. Operators are encouraged to take an integrated approach in operations and training (e.g., initial and recurrent) to emphasize the responsibilities and importance of PF and PM roles.

A critical aspect of monitoring duties includes intervention when a deviation is identified. Each operator's policies, procedures, and training should adequately cover flightpath intervention including human-to-human intervention.

Investigation Process

The investigation involved the National Transportation Safety Board (NTSB) of the United States of America as State of design and State of manufacturer, the Australian Transport Safety Bureau (ATSB) of Australia, the Transport Safety Investigation Bureau (TSIB) of Singapore as States providing assistance and Air Accident Investigation Bureau (AAIB) of Malaysia as State providing information that assigned accredited representatives according to ICAO Annex 13.

The Aviation Investigation Bureau (AIB) of Kingdom of Saudi Arabia acted as observer during the search of flight recorders.

Useful or Effective Investigation Techniques

The investigation was conducted in accordance with the KNKT approved policies and procedures, and in accordance with the standards and recommended practices of ICAO Annex 13.

KNKT share the experience of underwater flight recorder searching described in the Appendices of this report.

CHAPTER 7

ANALYSIS

The investigation found that there were several mechanical irregularities on the aircraft prior to the accident flight that had been previously reported. This analysis will discuss the reported problems and the rectifications performed. Since the accident, additional safety issues were identified that existed prior to the accident flight, such as basic aircraft design, aircraft certification, and organizational issues. The investigation did not find any issue related to the air traffic control transmissions and air traffic services radar system.

The analysis will discuss:

• Previous flight crew actions which discuss the LNI043 flight crew decision making that occurred during and after their flight.

• Maintenance in Jakarta which discusses the maintenance actions taken on the accident aircraft prior to the accident flight.

• Accident flight crew actions which discuss the relation between flight crew training and proficiency, flight crew awareness, flight crew workload, and flight crew resource management.

• Organizational issues which discusses the problems reported on previous flights, replacement of AOA sensor, hazard reporting, dispatch of aircraft, and maintenance management.

• MCAS design and certification which discusses the design of MCAS including the assessments, certification process, related regulations, and regulatory oversight.

- Xtra Aerospace's rectification process of the AOA sensor that was installed on the accident aircraft.

Previous Flight Crew (LNI043) Actions

Lion Air flight 043 (LNI043) was from DPS to CGK and was about 1.5 hours long. The LNI043 flight crew was able to successfully land the accident aircraft while experiencing the same conditions as the accident flight. The investigation looked at the flight crew's situation awareness and handling of flight deck indications, their decision to continue the flight, and their reporting of encountered issues after the flight.

Situation Awareness and Handling of Flight Deck Indications
The Captain's initial response, as the PF, to the activation of stick shaker during lift-off and subsequent response of numerous caution lights was to continue rotation by maintaining pitch 15 degrees and existing take-off thrust. After the Captain transferred control to the FO, he cross-checked the flight instruments and determined his instruments were erroneous. The Captain action of transferring the control prior to crosscheck of the instruments may have indicated that the Captain generally was aware of the repetitive previous problem of SPD and ALT flags and the replacement of the left AOA sensor on this aircraft.

During acceleration and clean up, there were three occasions where the aircraft did not climb positively. Following the advice from the deadheading crew on the observer seat that the aircraft was diving down, the Captain commanded the FO to follow F/D command and re-trim the aircraft to retain appropriate climb path. The FO commented that the aircraft is "too heavy to hold back" which suggests and the FO also unable to trim the aircraft as intended where the aircraft started to pitch down after nose-up trim was released. Observation of the aircraft to the condition reinforced the Captain to cut-out the Stabilizer Trim. This action made the aircraft under control and enabled the flight crew to fly the aircraft normally using the manual trim.

Decision to Continue the Flight
As the flight crew able to control the aircraft, the Captain declared urgency call (PAN PAN) to Denpasar Approach controller, reported their problem and performed three Non-Normal-Checklist of Airspeed Unreliable, Altitude Disagree and Runaway Stabilizer. During the execution of the non-normal-checklist, the Captain reengaged the Stabilizer Trim Cut-out switch, but it resulted in the aircraft to pitching down, consequently Stabilizer Trim Cut-out was returned to cut-out position and the flight crew resumed flying with manual trim. After completion of the non-

normal-checklist and discussion with the FO, the Captain decided to continue the flight despite the existing condition and the deadheading pilot advised him whether returning to the departure station would be appropriate. The stick shaker remained activated throughout the flight, the reliable PFD was on FO side, and aircraft was flown and trimmed manually. Further, the Captain of LNI043 felt confident to continue the flight to the destination because the aircraft was controllable and the expected weather along the route and at the destination was good.

The QRH states, "It is not possible to develop checklists for all conceivable situations and in some multiple failure situations, the flight crew may need to combine the elements of more than one checklist. In all situations, the captain must assess the situation and use good judgment to determine the safest course of action". The FCTM states that in a non-normal situation, the pilot-in-command, having the authority and responsibility for operation and safety of the flight, must make the decision to continue the flight as planned or divert. However, on this flight, the Captain's decision to continue to the destination was based on the fact that a requirement to "land-at-the-nearest-suitable-airport" in the three Non-Normal-Checklists was absent.

As the flight crew decided to continue the flight, they informed the Denpasar Approach controller that they had managed the situation, however when they contacted the Upper West Madura controller, they transmitted again the urgency call (PAN PAN) with additional information that they having instrument problems and requested to maintain flight level 280. This request indicated that the flight crew were aware with the existing aircraft system problems they would not be able to fly at an RVSM (Reduced Vertical Separation Minima) level. After being transferred to Upper West Semarang, the flight crew restated their problem with additional information and during the descent to destination they requested uninterrupted descent path profile. This action suggested that the flight crew were aware of their existing flight condition (continuous stick shaker, manual flying, manual trimming, FO PFD was the primary instrument) required a simplified flight path management until approach and landing.

The decision to continue a flight with stick shaker continually activated is not common. The Boeing records showed that reported events related to stick shaker activation during or shortly after takeoff on Boeing 737 aircraft had occurred 27 times between 2001 and 2018. Of those cases, the flight crew elected to return in 18 flights, three flights diverted to the nearest airport, five flights continued to their destination (including the LNI043 flight) and one flight unknown. The LNI043 flight crew decision to continue with stick shaker active is not common in comparison to previous events of erroneous stick shaker. When combined with the runaway stabilizer situation recognized by the flight crew, the decision to continue was highly unusual.

During the flight, due to the existing condition of the aircraft, the Captain told the FA1 that the fasten seat belt would be kept on all the way due to the weather condition. This indicated that the Captain sensed the situation was at higher risk and did not want the passengers to move around the aircraft while the flight crew was flying manually. Additionally, the Captain asked the deadheading pilot to assist, such as monitoring aircraft flight path, listening to the Air Traffic Controller (ATCo), ensuring that no checklist item was skipped, and calculating required Vref approach and N1 go-around values. This action indicated that the Captain was aware of the need to use all available resources to alleviate the matter to complete the flight to the destination, despite the increased workload and stressful situation.

Flight Crew Reporting/Documentation of Issues

After parking in Jakarta, the Captain made an entry in the AFML about problems experienced during the flight; the IAS DISAGREE and ALT DISAGREE warning and the FEEL DIFF PRESS light illuminated were reported. These three problems were the problems displayed in the messages on the PFD and the overhead flight controls panel. The Captain did not mention the activation of the stick shaker as he believed that the activation was the outcome of the IAS Disagree problem. The Captain also did not report the runaway stabilizer and the use of the STAB TRIM CUTOUT guarded switches or that he had to use manual trim for the majority of the flight and the landing. The incomplete report of the mechanical irregularities was not in accordance with the requirement of OM-part A, CASR Part 121 and ICAO Annex 6.

The Captain's incomplete report in the AFML was based on his incomplete understanding of the interrelationship between the effects experienced during the flight and the system failures that caused those effects, despite the fact they had isolated the problem after the STAB TRIM CUTOUT switches were moved to CUTOUT. Further, the requirement to report all known and suspected defects is very critical for engineering to be able to maintain the airworthiness of the aircraft.

In addition, the STAB TRIM CUTOUT switches were returned to normal after landing in Jakarta. Finding the STAB TRIM CUTOUT switches engaged would have been additional information to either the engineer performing maintenance.

Maintenance in Jakarta

The engineer in Jakarta reviewed the Captain's entry on the AFML which reported that IAS DISAGREE and ALT DISAGREE warning and FEEL DIFF PRESS lights illuminated on the previous flight. The fault code was not documented in the AFML. Thereafter, the engineer in Jakarta checked the OMF to focus the trouble shooting, however the engineer did not record the maintenance message that appeared in the OMF in the AFML. If the engineer was unaware of the maintenance message and the fault code, this would increase the difficulty for trouble shooting by the engineer.

The AFML showed that the engineer in Jakarta conducted the IFIM task for "ALT DISAGREE shows on PFD – captain's". However, the conduct of the IFIM task for "IAS DISAGREE shows on PFD – Captain's" was not recorded on the AFML.

The IFIM tasks of "ALT DISAGREE" on the step (3) of the "Fault Isolation Procedure" include the requirement to conduct the flushing by referring to the corresponding AMM task. This means that to complete the IFIM tasks, the flushing procedure on the AMM must be completed. The flushing procedure as mentioned on the AMM (which requires disconnecting the pitot and static lines) included a leak test of the pitot and static system which was described on the step K.

The IFIM task point B step (3) states that if the observed fault symptom has gone after the flushing, then the fault has been corrected. After conducting flushing of the left pitot ADM and static ADM, the engineer performed operational test and considered that the problem was solved. Thereafter the engineer released the aircraft to service. The leak test required by the AMM task on step K had not been conducted which mean that the IFIM task point B step (3) could not be considered as complete.

Both IFIM tasks of "ALT DISAGREE" and "IAS DISAGREE" describe similar tasks up to step (3) of the "Fault Isolation Procedure". The step (4) of the IFIM task of "ALT DISAGREE shows on PFD – Captain's", requires conducting left static system low-range leak test and left pitot system leak test. While the step (4) IFIM task for "IAS DISAGREE shows on PFD – Captain's" requires performing the right static low range leak test and right pitot system leak test. The leak test on both IFIM tasks refer to AMM task that is the same as the leak test in step (3). This indicated that the leak test required by the IFIM task step (4) was the repetition of the task on step (3). This repetition is inefficient and does not contribute to the problem solving.

Further step of the IFIM task of "ALT DISAGREE shows on PFD – Captain's" and "IAS DISAGREE shows on PFD – Captain's" requires performing some checks and performing visual inspection to the AOA sensors. There was no requirement to perform AOA value test. The IAS and ALT disagree reported which occurred on the LNI043 flight which was caused by AOA sensor bias, would not be able to solve by both IFIM tasks of "ALT DISAGREE shows on PFD – Captain's" and "IAS DISAGREE shows on PFD – Captain's".

AOA DISAGREE message was not enabled and was inhibited; therefore, it did not appear on the LNI043 flight. If the AOA sensor bias had generated the AOA DISAGREE message, the flight crew most likely would document the alert as it was displayed on the PFD. The AOA sensor bias would have been detected by the IFIM task "AOA DISAGREE" which requires checking the AOA values. The inhibited AOA DISAGREE message contributed to the inability of the engineer to rectify the failure of the AOA sensor.

The certified design of Boeing 737-8 (MAX) was to include an AOA DISAGREE message on all aircraft. The software which generates the AOA DISAGREE message was subcontracted by Boeing another company. The installed software did not include the AOA DISAGREE message for aircraft that was not installed with the AOA indicator. The Lion Air elected not to enable the AOA indicators on the PFDs and such the AOA DISAGREE message would not appear on both PFDs even though the DFDR recorded AOA value difference of about 21°.

The lack of an AOA DISAGREE message did not match the Boeing system description that was the basis for certifying the aircraft design. The software not having the intended functionality was not detected by Boeing nor the FAA during development and certification of the 737-8 MAX before the aircraft had entered service. Soon after, Boeing reviewed the situation and concluded that the inoperative AOA DISAGREE message on selected aircraft did not represent a safety of flight issue. One consideration was that additional maintenance alerts (e.g. stuck AOA or bent AOA) were still available. As a result, the implementation error was scheduled to be corrected for the next display system software update.

Accident Flight Crew Actions

The accident flight crew had flown together twice prior to the accident flight. Since the accident flight crew was presented with the same indications as the previous flight, the investigation looked at the flight crew's training and proficiency, the flight crew's workload, their awareness of the aircraft condition before the flight, and their crew resource management (CRM). Also reviewed was the Non-Normal Checklists (NNC) procedures, memory items, transfer of controls, flight crew communication, and the lack of MCAS training or flight crew awareness of MCAS.

Flight Crew Training and Proficiency

After the IAS DISAGREE had been identified, the Captain instructed the FO to perform memory items of Airspeed Unreliable, and the FO did not perform them. The first four items of the Airspeed Unreliable NNC are memory items to be performed by memory and must be done before reading the checklist. The Captain repeated the command about two minutes after without mentioning the NNC title and the FO was confused of the memory items to be performed. About 1 minute later, the FO asked to the Captain of the memory item to be performed to which the Captain responded "Airspeed Unreliable". The FO acknowledged and started to locate the checklist. About 1 minute later, the FO found the checklist and started to read the checklist.

The Airspeed Unreliable procedure is one of the checklists which are listed on the Quick Action Index. The Quick Action Index is available on the cover page of the QRH and the Airspeed Unreliable is on the second line of the list. The inability for the FO to perform memory items and locate the checklist in the QRH in a timely manner indicated that the FO was not familiar with the NNC. This condition was reappearance of misidentifying NNC which showed on the FO's training records.

The item number 8 of the Airspeed Unreliable NNC requires the flight crew to cross check the captain, first officer and standby airspeed indicators, when the aircraft is in trim and stabilized. The airspeed indication requires to be compared with a table and indication that differs by more than 20 knots or 0.03 Mach should be considered unreliable.

On the accident flight, the aircraft altitude was unable to be maintained and the MCAS activated repeatedly. These conditions did not meet the requirement of the NNC for the flight crew to compare the airspeed indicators and the table as the aircraft would not be in trim and stabilize condition. The FDR recorded that the differences of the Captain's and the FO's airspeed indicators was about 15 knots which was below the value to be considered as unreliable.

Despite the flight crew's attempt to execute of the NNC, due to increase workload, and distractions from the ATC communication, the NNC was unable to be completed in that situation. The unfinished NNC made it difficult for the flight crew of LNI610 to understand the aircraft problem and how to mitigate the problem.

While the aircraft control was being transferred by the Captain to the FO, the right altimeter indicated 5,900 feet, the pitch trim was 5.4 units. A few seconds later the MCAS activated for 8 seconds and the pitch trim changed from 5.4 to 3.4 units. The FO commented on abnormality of the aircraft control but did not clearly specify the abnormality. The FO commanded Aircraft Nose Up (ANU) trim and the pitch trim changed to 3.6 unit.

The pitch trim continued decreasing as the following activations of MCAS were not countered by the FO sufficiently trimming the aircraft nose up. The control column force increased up to 103 lbs (46 kg) while the aircraft still descended which indicated that the force exerted was insufficient to maintain aircraft altitude. During the repetitive MCAS activations, the Captain managed to control the aircraft altitude when the pitch trim was maintained above 5 units by commanding the ANU pitch trim to counter the MCAS trim down. The FO was unable to control the aircraft as the repetitive MCAS activations were not countered by adequate trim up input. The common flight crew reaction to a heavy control column is by providing adequate trim. This suggests that while the Captain's training or experience enabled him to recognize the need for sustained nose up trim, the FO's training and experience did not.

This condition was also in agreement with the FO's training records that showed several comments indicating that the FO had difficulty in aircraft handling. The Lion Air policy for such deficiencies was that the flight crew would be treated with additional briefings or rehearsal. The reappearance of difficulty in aircraft handling indicated that the treatment was not effective.

Flight Crew Workload

Before the flaps were retracted and MCAS activated, the workload for the flight crew was high but should have been manageable. Workload can be managed by training, knowing procedures, managing or taking control of the known situation and using crew resource management (CRM). During the takeoff roll, the DFDR recorded different values of left and right AOA sensors about 21°, but the difference was not displayed in the cockpit. The flight crew did not detect any aircraft abnormality. Just after liftoff, the left stick shaker activated and numerous messages on the PFD were displayed. After the flaps were retracted by the flight crew, automatic aircraft nose down trim was activated repeatedly which was triggered by MCAS. The MCAS activation required correction to maintain altitude by applying aft control column force to correct pitch and using electric trim to neutralize the column forces.

Typical markers for high workload include dropped task, reduced task performance, and reduced verbalization. High workload for the Captain could be identified by his short responses to the FO, his difficulty to maintain an assigned heading and altitude due to repetitive MCAS activation, and his failure to manage speed and thrust, and call out flap retraction points.

The FO asked the controller of the aircraft altitude and later also asked the indicated speed on the ATC radar display. The FO question may indicate that the flight crew were not confidence to the cockpit indications and attempted to obtain other source of information. However, the ATC radar receives altitude data transmitted by the aircraft's selected mode and the ATC radar system displayed the ground speed based on the aircraft movement calculation on the radar system therefore, no additional data may be acquired. Being unable to determine which source was unreliable might increase stress to the flight crew. The flight crew workload increased with the ATCo communication when the controller provided eight heading instructions after the flight crew reported that the aircraft was experiencing a flight control problem. The ATCo considered a flight control problem was not an emergency condition which was consistent with ATS SOP of AirNav Indonesia branch JATSC. There was also no objection by the flight crew to the ATCo heading instruction and the flight crew did not declare an emergency. The absence of a declaration of urgency (PAN PAN) or emergency (MAYDAY), or asking for special handling, resulted in the ATCo not prioritizing that flight. With priority, ATC would not require LNI610 to maneuver repeatedly.

During the simulation of the accident flight on the Boeing engineering simulator, the flight crew were distracted by the ATCo communication and claimed that it significantly increased the flight crew workload. The JATSC Standard Operation Procedure for Approach Control Services, Chapter 6.2, mentions that the conditions when flight crew report any instrument malfunction that might be suspected or classified as an emergency situation. The flight control problem as reported by the accident flight crew was not listed as a suspected emergency situation. The ATCo did not consider the flight control problem as an emergency situation that restricted communication.

Flight Crew Awareness of Condition of Aircraft
During the LNI610 flight preparation, the CVR recorded flight crew discussion related to the Deferred Maintenance Item (DMI) but the flight

crew did not discuss any issue related to previous aircraft problem recorded in the AFML. The OM-part A subchapter 2.1.14.3 describes that one of the Captain responsibilities is to examine the AFML to inquire about the technical status of the aircraft before the flight.

The absence of flight crew discussion of the previous problem suggests the flight crew might not be aware of aircraft problems that might reappear during their flight. This was different compared to the flight crew of the LNI043 flight, who had awareness of the aircraft condition after discussion with the engineer about the aircraft problem and the rectification prior to the flight which may have helped the flight crew to immediately identify the problem correctly.

Being unaware of multiple problems that occurred on the previous flight, including the stick shaker activation and uncommanded AND trim lead to the inability of the flight crew to predict and be prepared to mitigate the events that might occur.

In LNI610 flight, while the multiple problems occurred, the activation of stick shaker and MCAS activated repeatedly, the flight crew attempted to complete the NNC, as they are trained to refer to QRH in abnormal condition. The NNC should be performed to identify and mitigate the problem. However, the NNC was unable to be completed in that situation. Therefore, until the end of the flight, the flight crew was unable to complete the NNC and identify the problem. This made the flight crew did not understand the aircraft problem and how to mitigate. The activation of stick shaker indicated that the aircraft was about to stall while the cockpit instrument indicated the pitch was relatively level and the speed relatively high. The cockpit instrument did not indicate that the aircraft was close to stall condition which contradicted to the stick shaker activation. The aircraft was not equipped with AOA indicator and the AOA disagree message was inhibited, so there was no information provided to the flight crew of the AOA sensor that trigger the activation of stick shaker. This made the flight crew was not aware of the real aircraft condition.

Flight crews trained on previous versions of the B737 aircraft would have been aware of the AOA DISAGREE message on the PFD, should such a condition arise. However, because the AOA DISAGREE message was not available on B737-8 (MAX) aircraft not fitted with the optional AOA indicator, flight crews would not be aware that this message would not appear if the AOA DISAGREE conditions were met. This would contribute to flight crew being denied valid information about abnormal conditions being faced and lead to a significant reduction in situational awareness by the flight crew.

No information about MCAS was given in the flight crew manuals and MCAS was not included in the flight crew training. These made the flight crew unaware of the MCAS and its effects. There were no procedures for mitigation in response to erroneous AOA.

Crew Resource Management

Crew Resource Management (CRM) refers to the effective use of all available resources: human resources, hardware, and information. These activities include communication, problem solving, decision making, maintaining situation awareness, workload management, and dealing with automated systems.

During the accident flight just after airborne, the left stick shaker activated, followed by IAS DISAGREE message. Shortly after, the Captain exclaimed that the aircraft had problems. The FO mentioned auto brake disarm twice and IAS disagree, and then asked whether the Captain intended to return.

The guidelines for handling a non-normal situation, according to the Non-Normal Operation chapter in the FCTM, are recognition of non-normal condition, maintain aircraft control, analyze the situation, take the proper action, and evaluate the need to land.

The Lion Air OM-part A describes that during an abnormal situation, "the PIC must allocate the crew duties to ensure that the highest level of situation awareness is maintained in the cockpit and cabin. This will prevent all attention being totally directed at resolving the emergency or abnormal situation to the detriment of safe flight. Any ambiguities, confusion, unresolved discrepancies or use of improper procedures must be discussed immediately, and if necessary, a missed approach initiated to allow remedial action at safe altitude." The Captain did not verbalize allocation of the crew duties and did not clearly verbalize the plan to hold or to troubleshoot the aircraft problems.

The Captain and FO did not have a shared mental model of the situation as exhibited by their lack of clear and effective communication. For example, the FO did not conduct memory items for the airspeed unreliable NNC when asked by the Captain and the Captain did not verify that the FO did not conduct the memory items when asked. The Captain also did not verbalize information regarding the aircraft state or the need to trim out increased column forces when he transferred control to the FO near the end of the flight.

The Lion Air OM-part A describes that during abnormal and emergency situations, ATC communication is the task of PF which in this flight was the Captain, while completing procedure is the task of PM. During this flight, the ATC communication and completing checklist were handled by the FO as PM.

The CVR showed that, while the FO reading the NNC, the Captain responded with short answer or did not respond. Meanwhile, when the aircraft heading or altitude was not correct, the FO did not provide the required callouts to the Captain except one altitude callout. The Captain also did not call the flap positions to the FO and had to be prompted by the FO after passing the flap retraction speeds. This might be due to both pilots focusing on their own task. The Captain was focusing on controlling the aircraft which had flight control problem and the FO was focusing on completing the procedure which took time for the FO to accomplish. Additionally, the FO was not monitoring the flight path progress as he was performing an NNC procedure. Both pilots being preoccupied with individual tasks indicated that the crew coordination was not well performed.

The DFDR recorded that the Captain managed to control the aircraft altitude by counteracting the MCAS action. The Captain did not verbalize to the FO the difficulty in controlling the aircraft and the need for repeated aircraft nose up trim. The FO was preoccupied with completing the NNC and not monitoring the flight progress. It could not be determined why the Captain did not communicate this critical piece of information: whether he was unaware that he was trimming to the extent he was, that high workload led him to ask shed the importance of communicating the state of the aircraft, or that he believed the FO would naturally recognize the need to trim the aircraft without him having to verbalize this.

Subsequently, the FO was not able to anticipate the need for repeated pitch corrections and nose up trim and did not use adequate electric trim to counter repetitive MCAS activations while controlling the aircraft.

The Lion Air OM-part A related to handover control did not mention any requirement to describe specific handling situation to the flight crew receiving the control. Even though it was not required per Lion Air procedure or Indonesia requirement, the absence of Captain's specific description contributed to the FO's difficulty to understand the situation and may have contributed to his inability to mitigate the problem.

The Captain training record showed a remark of Crew Resource Management (CRM) need to be improved and one assessment on "teamwork exercise". The remark was to use standard signal for effective communication and good team work during abnormal or emergency situation.

The flight crew, as a team, should have a common goal of flying the aircraft safely. The essence of team action is the coordination, synchronization, and integration of the contributions of all team members into a series of collective responses. The required action by the flight crew as a team is described in the procedure, while in this accident flight the procedure of ATC communication task was not implemented. The flight crew was preoccupied on their own tasks without coordination. Most of the components of effective crew coordination were not achieved, resulting in failure to achieve the common goal of flying the aircraft safely.

Organizational Issue

Problems Handling on Previous Flights

The first indication of the SPD (speed) and ALT (altitude) flags on the Captain's PFD was recorded when the aircraft arrived at Manado from Tianjin, China, on 26 October 2018. The maintenance actions taken by the engineer in Manado were believed to have rectified the aircraft problem and the engineer then released the aircraft to service.

The aircraft then continued the flight from Manado to Denpasar. The DFDR recorded similar signatures of faulty data on left Calibrated Air Speed (CAS) and barometric altimeter during cruising which indicated that the SPD and ALT flags appeared on the Captain's PFD. The event of SPD and ALT flag on Captain's PFD was the second event on the aircraft and according to the CMM should be classified as repetitive problem.

Subsequently, the aircraft continued the flight from Denpasar to Lombok and returned. The DFDR showed valid AOA values which indicated that the airspeed indicator and altimeter functioned normally. This was consistent with the result of AOA sensor examination which indicated that the resolver 2 became unreliable during cold temperature, and no cold temperatures were encountered on the relatively low altitude flight to and from Lombok.

The aircraft continued the flight from Denpasar to Manado. After that flight, the flight crew reported on the AFML that the SPD and ALT flag appeared on Captain's PFD. The DFDR recorded the signature of faulty data on left CAS and Barometric Altitude during cruise until landing.

The AFML recorded that referring to IFIM task 27-32-00-810-816 (SMYD Fault) the engineer in Manado performed a self-test of the SMYD 1 with result fail. The correlated maintenance messages shown on the OMF directed the engineer to conduct Build in Test Equipment (BITE) test on the Flight Management Computer - Control Display Unit (FMC CDU) which showed the Angle of Attack (AOA) signal fail. The engineer aware that to complete rectification required replacement of left AOA sensor and did not have the spare part available in Manado. The IFIM task also ordered to perform wiring check of the left Air Data and Inertia Reference Unit (ADIRU) however, due to rain, the engineer did not perform wiring check due to the lightning hazard.

Referring to the IFIM, the engineer reset the circuit breaker (CB) of the left ADIRU and SMYD 1. After resetting the CB, the engineer conducted the system test of SMYD 1 and Digital Flight Control System (DFCS) test via FMC CDU to which the system passed.

The following morning during flight preparation, the engineer met with the flight crew and described the rectification of SPD and ALT flag problems. The flight crew that had flown the aircraft from Manado to Denpasar were same flight crew who flew from Denpasar to Manado. The flight crew mentioned that the problems appearing several times and should be considered as repetitive trouble. The engineer mentioned that problem rectification would be better to be performed in Denpasar and suggested to continue the flight considering that the SPD and ALT flags had no longer appeared on the Captain's PFD. This indicated that the aircraft was released with known possible recurring problem.

In the flight from Manado to Denpasar, the flight crew reported that during takeoff roll the Auto-throttle (A/T) disengaged and while passing 7,000 feet the SPD and ALT flags appeared on Captain's PFD. The DFDR recorded the A/T disengaged on takeoff roll and the signatures of faulty data on left CAS and Barometric Altitude occurred after both engines started. This means that the SPD and ALT flags on the Captain's PFD most likely had appeared after the engine start. The disengagement of A/T during takeoff roll was also caused by the failure of the Captain-side ADIRU speed input to the A/T system.

The Lion Air OM-part A stated that any failure that occurs after off block requires accomplishment of abnormal procedure. After the failure has been identified and prior to takeoff, the Minimum Equipment List (MEL) must be consulted and if the MEL classified as a "NO GO" item, the flight crew must return. The Boeing 737-8 (MAX) MEL did not include indicated airspeed or altimeter. An item that is not included in the MEL and related to airworthiness is required to be operative. The decision for the flight crew continuing the flight with an unserviceable airspeed indicator and altimeter was contrary to the company procedure.

The disengagement of A/T during takeoff and the DFDR data showed the airspeed indicator and altimeter had failed prior to takeoff should have prevented the aircraft for dispatch, however, the flight crew elected to continue the flight which contrary to the Lion Air manual.

Replacement of AOA Sensor

After arriving in Denpasar at 0205 UTC (1005 LT), the flight crew reported that SPD and ALT flags appeared on Captain's PFD, and the Auto Throttle disengaged on takeoff roll. The flight crew mentioned to the engineer in Denpasar that the problems had appeared repeatedly.

The engineer checked the OMF and found the maintenance messages of 27-31012 "AD data invalid", 27-31015, "ADIRU-L inertial data is invalid" and 34-21107 "ADIRU-L ADR data signal is invalid" and 34-21123 "AOA signal out of range". The engineer conducted BITE test via FMC CDU and found maintenance messages of 34-21007 "ADR Data Invalid" and 34-21023 "AOA signal fail".

The maintenance message of 34-21123 "AOA signal out of range" lead to IFIM task 34-21-00-810-828 "AOA signal is out of range".

The IFIM tasks "AOA signal is out of range" on step C (5) directed to replace the AOA sensor if the maintenance message was still active. The engineer in Denpasar considered that the problem had appeared repeatedly and for the trouble shooting, the engineer decided to replace the left AOA sensor. The engineer coordinated with Maintenance Control Centre (MCC) in Jakarta, requesting AOA sensor spare.

The AOA sensor spare arrived in Denpasar about 1000 UTC (1800 LT) and the engineer replaced the left AOA sensor.

After the installation, the AMM required an installation test that can be performed with recommended method which requires test equipment of AOA test fixture SPL-1917 or alternative method using the SMYD BITE module. The test fixture was not available in Denpasar therefore the engineer performed the installation test by alternative method.

The alternative method was performed by deflecting the AOA vane to fully up, center, and fully down while observing the indication on the SMYD computer for each position. The AOA values indicated on the SMYD computer during the test were not recorded even though BAT procedures required it, however the engineer in Denpasar stated that the test result was satisfactory. The rectification was completed about 1230 UTC (2030 LT).

The engineer in Denpasar provided to the investigation some photos of the SMYD unit during an installation test as evidence of a satisfactory installation test result. The investigation confirmed that the SMYD photos were not of accident aircraft and considered that the photos were not valid evidence.

Following the accident, NTSB and Boeing performed AOA sensor installation test on an aircraft with an AOA sensor that deliberately misaligned by 33° bias before install. The installation test was performed by alternative method referred to in the AMM. The test result indicated that a misaligned AOA sensor would not pass the installation test as the AOA values shown on the SMYD computer were out of tolerance and "AOA SENSR INVALID" message appeared in the SMYD BITE module. This test verified that the alternate method of the installation test could identify a 21° bias in the AOA sensor.

After the installation test, the Denpasar engineer considered the aircraft problem had been solved and released the aircraft for flight. The Denpasar engineer briefed the LNI043 flight crew about the aircraft problems and the rectification that had been performed. The Captain was convinced that the problems had been solved.

On the flight from Denpasar to Jakarta, the DFDR recorded AOA differences with the value in the left AOA sensor was approximately 21° higher than the right AOA sensor. Referring to the Boeing AMM for AOA installation test result, the misaligned AOA sensor should have been detected during the installation test using alternative method in Denpasar.

Comparing the results of the installation test in Denpasar and Boeing, the investigation could not determine that the AOA sensor installation test conducted in Denpasar were successful.

Replacement of AOA sensor proved to be the solution to rectify the SPD and ALT flags that were reported appeared on Captain's PFD, however the AOA sensor installed was misaligned by about 21° and the result of the installation test in Denpasar could not be determined with any certainty.

Hazard Report

The reporting procedure of hazard or occurrence described in the OM-part A was different with the description on the SMSM. In addition, the web-based application software mentioned in both documents referred to the old web-based application which had been changed, while the initial SMS training material revision 00-2019 described the new web-based application software. The LNI043 Captain reported the flight event to the SS Directorate using the new web-based application software by filing the ASR online form.

The different descriptions in several Lion Air different between Lion Air CMM and BAT AMOQSM r company manuals indicated that the company manuals were not synchronized.

The Lion Air SMS training syllabus included the risk management subjects consisted of hazard identification and three other topics which scheduled to be delivered in 1 hour. The syllabus also include 1 hour group activity of hazard observation and risk managements. The SMS training recurrent was conducted every 3 years and delivered in 2 hours duration. The recurrent includes the risk management subject which delivered in 30 minutes.

The investigation considered that the amount of time to cover the hazard identification topic in the SMS training syllabus was insufficient. This may reduce the ability of employees to define and report a hazard. Consistently, the Lion Air safety report on December 2018 mostly consisted of occurrence report and only about 5 percent of hazard report.

The LNI043 flight that experienced multiple malfunctions were considered caused or could have caused difficulties in controlling the aircraft. According to the ICAO Annex 13, CASR part 830 and OM-part A, the flight is classified as serious incident which required investigation by the KNKT in accordance with the Aviation Law Number 1 of 2009 and Government Decree Number 62 of 2013.

The OCD and Duty Management Pilot as part of Operation Directorate handle operational issues after normal office hour and responsible to ensure the Safety and Security (SS) Directorate is notified when serious incident occurs after the normal office hour.

The problems that occurred on flight LNI043 was reported by the Captain to the SS Directorate and Duty Management Pilot. Specifically, the Captain mentioned that the flight experienced airspeed disagree and altitude disagree. The report did not mention information of stick shaker activation, Stabilizer Trim Cut Out guarded switches selection and the difficulties to control the aircraft. With the information provided, the Duty Management Pilot considered that the occurrence a daily aircraft problem and it was not classified as urgent safety attention, critical high severity or airworthiness matters of concern. Therefore, the Duty Management Pilot did not seek the detail of the event to the LNI043 Captain nor coordinated with the SS Directorate. The assessment to the report might have the same result if the same report was also reported to the OCD. The AFML entry for LNI043 which did not contain additional details about what was experienced was not in accordance with company guidance provided in OM-Part A, Section 11.4.9 which lists reportable events to include "Warning or alert, including flight control warnings, door warnings, stall warning (stick-shaker), fire/smoke/fumes warning".

The SS Directorate did not notice the occurrence since the report was filed outside normal office hours and the report to Safety and Security Directorate was not processed until the office hours on the following day.

The insufficient SMS training and inability of the employees to identify the hazard might also be indicated by the incomplete post-flight report of the problems that occurred on LNI043. The incomplete report became a hazard as the known or suspected defects were not reported which might make the engineer unable to properly maintain the airworthiness of the aircraft.

Content of the report did not trigger the Duty Management Pilot to assess this as a Serious Incident and enable a safety investigation. The risk of the problems that occurred on the flight LNI043 were not assessed to be considered as a hazard on the subsequent flight.

Maintenance Management

The maintenance management of the Lion Air was the responsibility of the Maintenance and Engineering Directorate. The maintenance program of the Boeing 737-8 (MAX) was described in the Lion Air Continuous Airworthiness Maintenance Program (CAMP) Boeing 737-8 (MAX). The investigation found that the CAMP applicable on the day of the accident was revision dated 28 February 2018. The PK-LQP was operated by Lion

Air since August 2018 had been included in the ACL and temporary revision of aircraft applicability list of the CAMP. The ACL referred to the CAMP issued on 28 February 2018 and investigation did not receive the ACL revision to reflect the CAMP temporary revision. This indicated that the CAMP was not updated in timely manner.

The maintenance performance of the Lion Air fleet was conducted by the Batam Aero Technic (BAT) under the agreement which covered the aircraft maintenance activity. The agreement does not relieve the Lion Air of their aircraft airworthiness responsibility, refer to the CASR part 121 subpart 121.363. This policy was described in the Company Maintenance Manual (CMM) where the maintenance audit and surveillance were conducted by the Lion Air quality assurance department and the maintenance oversight conducted by the Lion Air quality control department. The Lion Air Fleet Maintenance Management (FMM) department monitors the aircraft serviceability daily.

The FMM department coordinated with the BAT Maintenance Control Centre (MCC) department and BAT "Bapak Asuh" for daily aircraft serviceability monitoring including the aircraft repetitive problem. The duties and responsibilities of the MCC department and the BAT "Bapak Asuh" have not been included in the BAT AMOQSM procedure manual.

The definition of an aircraft repetitive problem was different between Lion Air CMM and BAT AMOQSM. The Lion Air CMM described that the aircraft problem categorized as the repetitive problem if discrepancy twice recurs on the same aircraft during 30 consecutive days of operation, while BAT AMOQSM stated three times within 30 consecutive days. This difference indicated that the Lion Air did not monitor the repetitive problem policy of the BAT as a subcontracted entity.

The AFML is the only source of the daily aircraft problem monitoring in which the problem may be identified by the flight crew or engineer. If the aircraft problem is not stated in the AFML, the repetitive problem may not be detected. The investigation found that the SPD and ALT flags problem was reported twice in the AFML on 26 and 27 October 2018 while the DFDR recorded the problems occurred three times. The SPD and ALT flags problem during the flight from Manado to Denpasar on 27 October 2018 was recorded on the DFDR but was not reported in the AFML. The absence of aircraft problem report affected the repetitive problem identification. The investigation did not find any evidence of handling the problem as repetitive according to the CMM, other than the statement on the AFML for replacement AOA sensor was "due to repetitive problem".

The OMF has the history page which contains record of the aircraft problems which can be utilized as a source for aircraft problem monitoring. The BAT has not utilized the OMF information as the source of aircraft problem monitoring.

During the replacement of the left AOA sensor, the installation test of the AOA sensor required the engineer to check the angle deflection of AOA sensor via the SMYD. The BAT LMPM required the engineer to record the test values to ensure that the test results were within tolerance. The engineer did not record the value of the AOA angle deflection during the AOA sensor installation test.

The OMF and Interactive Fault Isolation Manual (IFIM) provide trouble shooting guidance for the engineer. The investigation found that the engineers were prone to entering the problem symptom reported by the flight crew in the IFIM first instead of reviewing the OMF maintenance message. Conducting this method might lead the engineers into the inappropriate rectification task.

The Fault Reporting Manual (FRM) helps to directly appoint the proper IFIM task by fault code for particular problem entered by the flight crew. The fault code may direct the engineer to the relevant problem and prevent the unnecessary presentation of several faults, IFIM tasks or maintenance messages. The investigation found that all AFML pages received by the investigation did not contain fault codes. The absence of the fault code reported by the flight crew may increase the workload of the engineer and prolong the rectification process.

MCAS Certification

Design/Certification

Reason MCAS added to Aircraft
The Boeing 737-8 (MAX) is derivative of the 737-800 model and is part of the 737 MAX family (737 MAX 7, 8, and 9). The Boeing 737-8 (MAX) incorporated the CFM LEAP-1B engine, which has a larger fan diameter and redesigned engine nacelle compared to engines installed on the 737 Next Generation (NG) family. Because the 737-8 is a derivative of the 737-800 model, its certification basis, which was established per 14 CFR 21.101 Changed Product Rule, required Boeing to demonstrate compliance with Amendment 25-136 for significant areas of change at the product level and those areas affected by the significant product level change.

During the preliminary design stage of the Boeing 737-8 (MAX), Boeing tests and analysis revealed that the addition of the LEAP-1B engine and associated nacelle changes was deemed likely to negatively affect the stick force per g (FS/g) characteristics required by 14 FAR 25.255 and the controllability and maneuverability requirements of 14 FAR 25.143(f). After the study of various options for addressing this issue, Boeing implemented aerodynamic changes as well as a stability augmentation function called the Maneuvering Characteristics Augmentation System (MCAS), as an extension of the existing Speed Trim System (STS), to improve aircraft handling characteristics at elevated angles of attack. The MCAS was needed in order to make the Boeing 737-8 (MAX) handling characteristics so similar to the NG versions that no simulator training was needed for type rating. It was also required so that the 737 MAX passed the certification that the pitch controls could not get lighter on the approach to stall. If the aircraft had substantially different pitch behavior, then there would be a simulator training requirement for the pilots.

As the development of the Boeing 737-8 (MAX) progressed, the MCAS function was expanded to low Mach numbers. MCAS is designed to function only during manual flight (autopilot not engaged), with the aircraft flaps up, at an elevated AOA.

FHA for MCAS Related Failures

The investigation reviewed sections of Boeing's 737 NG/MAX Stabilizer Trim Control System Safety Analysis that pertained to MCAS. Boeing's analysis included a summary of the functional hazard assessment findings for the 737 MAX stabilizer trim control system. For the normal flight envelope, Boeing identified and classified two hazards associated with "uncommanded MCAS" activation as "major". The major classification used by Boeing indicated a remote probability of this hazard occurring and that it could result in reduced control capability, reduced system redundancy, or increased crew workload. Other classification categories include "Minor," "Hazardous," and "Catastrophic." Because uncommanded MCAS function was considered "Major," Boeing did not perform a specific fault tree analysis for an uncommanded MCAS hazard.

One of these hazards, applicable to the MCAS function seen in this accident, included uncommanded MCAS operation to original maximum authority of 0.6°. Boeing indicated that, as part of the functional hazard assessment development, flight crew assessments of MCAS-related hazards were conducted in an engineering flight simulator with motion capability, including the uncommanded MCAS operation (stabilizer runaway) to the MCAS maximum authority.

This assessment of Major failure effect did not require Boeing to more rigorously analyze the failure condition in the safety analysis, using Failure Modes and Effects Analysis (FMEA) and Fault Tree Analysis (FTA), as these are only required for Hazardous or Catastrophic failure conditions.

During the process of developing and validating the Functional Hazard Analysis (FHA), Boeing considered four failure scenarios including uncommanded MCAS function to the maximum authority limit of 2.5° of stabilizer movement. However, the uncommanded MCAS function to maximum authority was only flight simulated to high speed maximum limit of 0.6°, but not to low speed maximum limit of 2.5° of stabilizer movement. Boeing also not considered repetitive erroneous MCAS activations.

To perform the simulator tests, Boeing induced a stabilizer trim input that would simulate the stabilizer moving at a rate and duration consistent with the MCAS function. During this investigation, Boeing indicated to investigators that the failure modes that could lead to uncommanded MCAS activation (such as an erroneous high AOA input to the MCAS, as occurred in the Lion Air accident) was not simulated as part of these functional hazard assessment validation tests.

Boeing also indicated that engineering and test pilots discussed the scenario of repeated uncommanded MCAS activation during development of the Boeing 737-8 (MAX) and deemed it no worse than single uncommanded MCAS activation because it was assumed that the pilots would trim out uncommanded trim inputs to maintain control of the aircraft.

Repetitive MCAS activations without adequate trim reaction by the flight crew would escalate the workload and hence failure effects should have been reconsidered. During FHA, the simulator test had never considered a scenario in which the MCAS activation allowed the stabilizer movement to reach the maximum MCAS command limit of 2.5° of stabilizer movement. Therefore, their combined flight deck effects were not evaluated.

The FHA for uncommanded MCAS activation was classified as Major therefore, the FMEA and FTA were not required. FMEA would have been able to identify single-point and latent failures which have significant effects as in the case of MCAS design. It also provides significant insight into means for detecting identified failures, flight crew impact on resolution of failure effect, maintenance impact on isolation of failure and corresponding restitution of system.

FTA would have also been able to show if the system architecture meets the numerical criteria set by the FHA. Again, in general, only failures categorized as Hazardous or Catastrophic are evaluated, even though in some situations, complex single-string Major failures are evaluated. Another benefit of FTA that had been missed was to demonstrate compliance with probabilities for combinations of failures. If a system does not meet minimum allowable probability, FTA can indicate where system is deficient and where mitigating action can be applied.

Assumptions on Flight Crew Initial Response Time

In performing the FHA, Boeing considered the following factors and assumptions consistent with 14 CFR 25.671, 25.672, and guidance from AC25-7C for 25.143:

• Uncommanded stabilizer motion, regardless of source, is readily recognizable to the flight crew via unintended flight path changes and unexpected motion of the trim wheel.

• The MCAS stabilizer motion is limited to the maximum MCAS authority without flight crew action and the flight path can be controlled with column input alone without excessive short-term force.

• The crew will take prompt action to reduce the high column forces and return the aircraft to trimmed flight.

• It is appropriate to rely on trained crew procedures, especially if these procedures are considered memory items.

The assessment was also based on an assumption that the flight crew was highly reliable to respond correctly and in time. Boeing followed FAA guidance that the flight crew would respond within 3 seconds to any changes in flight condition. The assessment was that each MCAS input could be controlled with control column alone and subsequently re-trimmed to zero column force while maintaining flight path.

Following the accident Boeing issued flight crew Operation Manual Bulletin (OMB) that emphasize the procedure of runaway stabilizer in the case of uncommanded nose down stabilizer trim due to erroneous AOA. This was a trained flight crew memory item in 737-NG which could have been an option available to the flight crew. However, the procedure was not reintroduced during transition training and there was no immediate indication available to the flight crew to be able to directly correlate the uncommanded nose down stabilizer to the procedure. Therefore, the assumption of relying on trained crew procedures, to implement memory items was inappropriate.

For the first MCAS activation in the accident flight, the Captain responded less than 2 seconds, with aft control column but MCAS continued to move the horizontal stabilizer AND for a total of 10 seconds, after which the Captain applied electric trim. At the second MCAS activation, the Captain responded with electric trim after 8 seconds. At subsequent activations, the Captain responded after more than 3 seconds. For the previous LNI043 flight, the Captain responded in 2-3 seconds after the first MCAS activation. It should be noted that the flight crew did not react to MCAS activation but to the increasing force on the control column. Since the flight crew initially countered the MCAS command using control column, the longer response time for making electric stabilizer trim inputs was understandable.

During the accident and previous LNI043 flights, the flight crew initially responded in the same way, by pulling back on the control column. However, they did not consistently trim out the resulting column forces as had been assumed. As a result, Boeing assumption was different from the flight crew behavior in responding to MCAS activation.

Assumptions on Trimming out Forces Resulting from Flight Control Failures

14 FAR 25.255 stipulates that the aircraft must have satisfactory maneuvering stability and controllability with the degree of out-of-trim in both the aircraft nose-up and nose-down directions, which results from the greater of a 3-seconds movement of the longitudinal trim system at its normal rate for the particular flight condition with no aerodynamic load, except as limited by stops in the trim system, or the maximum mis-trim that can be sustained by the autopilot while maintaining level flight in the high speed cruising condition.

At its normal rate, after 3 seconds, the MCAS would move the stabilizer 0.81 degrees in the aircraft nose down direction. In response, the flight crew would be expected to maintain control of the aircraft using the controls available to them. Using the control column alone, the flight crew can maintain control using only aft column force. If the flight crew uses manual electric trim, it will allow the flight crew to control the aircraft by moving the stabilizer, but it will also reset the MCAS function which can be activated again in 5 seconds.

In the event of repetitive MCAS activation with repeated electric trim inputs by flight crew, but without sufficient flight crew response to return the aircraft to a trimmed state, the control column force to maintain level flight could eventually increase to a level where control forces alone may not be adequate to control the aircraft.

During the accident flight, the DFDR recorded a control force of 103 lbs., after repetitive MCAS activation was responded with the FO had responded with inadequate trim to counter MCAS. At this point, the flight crew was unable to maintain altitude.

Assumptions on Timing of Crew Recognition of and Reaction to Flight Control Failures

14 FAR 25.1329(g) states that under any condition of flight appropriate to its use, the flight guidance system may not produce hazardous loads on the aircraft, nor create hazardous deviations in the flight path. This applies to both fault-free operation and in the event of a malfunction, and it assumes that the flight crew begins corrective action within a reasonable period of time. The safety assessment described in Chapter 8 of AC 25.1329 establishes the failure condition for which appropriate testing should be undertaken. Assessment of failure conditions shall consider flight crew recognition of the effects of the failure condition, flight crew reaction time that is, the time between flight crew recognition of the failure condition and initiation of the recovery, and flight crew recovery.

The flight crew may detect a failure condition through aircraft motion cues or by cockpit flight instruments and alerts. The recognition time should not normally be less than 1 second. The flight crew reaction time is considered dependent upon the flight crew attentiveness, based upon the phase of flight and associated duties. During climb, cruise, descent, and holding, recovery action should not be initiated until at least 3 seconds after the recognition point. The recovery action should be commenced after the reaction time. Following such delay, the flight crew should be able to return the aircraft to its normal flight attitude under full manual control without engaging in any dangerous maneuvers during recovery and without control forces exceeding 75 lbs. for short term application with both hands on the control column or 10 lbs. for long term application as required in 14 FAR 25.143(d).

During the accident, multiple alerts and indications occurred which increased flightcrew's workload. This obscured the problem and the flight crew could not arrive at a solution during the initial or subsequent automatic AND stabilizer trim input, such as performing the runaway stabilizer procedure or continuing to use electric trim to reduce column forces and maintain level flight.

Without prior knowledge of MCAS functions, the flight crew would depend on the visual and motion cues, prior training for runaway stabilizer, and general training on pitch control to be able to analyze the situation and recognize the non-normal condition. Review of the DFDR data showed that during both the accident and the previous LNI043 flights, the flight crew responded within 2-3 seconds using control column to control the flight path and subsequently trimmed out column forces using electric trim. In the previous LNI043 flight, the flight crew required 3 minutes and 40 seconds rather than seconds to recognize and understand the problem, during which repetitive uncommanded MCAS activations occurred. During the accident flight, recognition of the uncommanded stabilizer movement as a runaway stabilizer condition did not occur thereby, the execution of the non-normal procedure did not occur.

Assessment of Flight Crew Workload in the Event of Erroneous MCAS Activation

During the Functional Hazard Analysis (FHA), unintended MCAS-commanded stabilizer movement was considered a failure condition with Major effect in the normal flight envelope. Boeing reasoned that such a failure could be countered by using elevator alone. In addition, stabilizer trim is available to offload column forces, and stabilizer cutouts is also available but not required to counter failure. As stated earlier, the events that led to the accident flight showed that in the event of repetitive MCAS activation without sufficient trim commands to return to trimmed flight, the cumulative mis-trim could not be countered by using elevator alone.

When the MCAS activates, it automatically commands an aircraft nose-down (AND) input to the horizontal stabilizer. If uncommanded MCAS is activated and the aircraft pitches down, causing the aircraft to descend, the flight crew can pull back on the control column to raise the nose. If the flight crew uses manual electric trim, the MCAS resets. That means 5 seconds, the system can make another AND command requiring the flight crew to respond again.

In the event of an MCAS activation with manual electric trim inputs by the flight crew, the MCAS function will reset which can lead to subsequent MCAS activations. With an MCAS command due to an erroneous high AOA signal, and flight crew inputs that do not fully return the aircraft to a trimmed state, subsequent MCAS commands can result in the aircraft becoming significantly miss-trimmed.

To recover, the flight crew could: 1) stop making manual electric trim inputs (which would stop resetting MCAS), or 2) make sufficient (long duration) manual electric trim inputs to return the aircraft to a trimmed state, or 3) use the stabilizer trim cutout procedure to stop all electric trim commands and control trim using the manual trim wheels. If one of these three flight crew trim responses is not used, a subsequent combination of

MCAS commands and short-duration (insufficient) flight crew electric trim inputs may result in a mis-trimmed condition that cannot be controlled.

A combination of repetitive MCAS-commanded coupled with flight crew electric trim input led to a flight condition that considerably increased the flight crew workload of maintaining control. The previous LNI043 flight showed that repetitive MCAS-commanded stabilizer movement was able to be countered by the flight crew by repeatedly trimming out erroneous aircraft nose down trim (option 2 noted above) and was only able to be stopped by Stabilizer Trim Cutout switches (option 3 noted above), enabling the flight crew to safely continue flight and land in Jakarta. Boeing reasoning that the stabilizer cutout is available but not required is incorrect.

During the FHA, Boeing did not adequately assess the effect of repetitive MCAS activation. The repetitive MCAS-commands coupled with insufficient flight crew electric trim inputs, may have led to increasing flight crew workload.

Single and Multiple Failure Analysis

14 FAR 25.671 (c) states that "probable malfunctions must have only minor effects on control system operation and must be capable of being readily counteracted by the pilot". This includes any single failure, such as the AOA sensor malfunction.

Boeing performed a single and multiple failure analysis to help validate system functional hazard assessments (FHAs), design assurance level (DAL) assertions, and extended operations (ETOPs). The analysis considered the air data system's worst-case scenario of "Erroneous Left & Right Air Data" and "Loss of one AOA followed by Erroneous AOA" (both dual failures). The air data systems worst case scenario would be that the three altitudes and airspeed display's showed different values from each other. This unreliable airspeed/altitude failure effect was deemed potentially catastrophic prior to flight crew recognition of the issue. Boeing considered that the loss of one AOA and erroneous AOA as two independent events with distinct probabilities. The combined failure event probability was assessed as beyond extremely improbable, hence complying with the safety requirements for the Air Data System. However, the design of MCAS relying on input from a single AOA sensor, made this Flight Control System susceptible to a single failure of AOA malfunction.

During the accident flight, the scenario was initiated by a single failure, a high bias in AOA sensor. This high bias resulted in several aircraft level effects including stick shaker, erroneous airspeed and altitude displays and MCAS after the flaps were retracted.

Over the last 17 years in the Boeing 737 fleet alone, there were 25 events of stick shaker activation during or shortly after takeoff. Because of these events, Boeing should have foreseen and assessed AOA sensor failures. The FCC controlling the MCAS is dependent on a single AOA source, the MCAS contribution to cumulative AOA effects should have been assessed.

During the single and multiple failure analysis from the air data system worst case scenario of "failure of one AOA followed by erroneous AOA", Boeing concluded that the effect would be hazardous until the flight crew recognized the problem and took appropriate action to mitigate it. Boeing recognized that training would improve flight crew recognition and response. The Airspeed Unreliable NNC and the Runaway Stabilizer NNC provide guidance and training for recognition and response. However, the accident flight crew was unable to complete the Airspeed Unreliable NNC before the accident occurred nor were they able to recognize the need for the runaway stabilizer NNC. Since the training or the guidance for actions taken in such situation were not provided, the effect category should have remained hazardous.

Decision to Rely on Single Sensor

The Boeing 737-8 (MAX) is equipped with two Flight Control Computers (FCCs) and two Angle of Attack (AOA) sensors. The Speed Trim System (STS), including the MCAS function, is a flight control law contained within each of the two FCCs. STS is only active in the master FCC for that flight. At aircraft power-up, the master FCC defaults to the left side FCC; and will then alternate between the left and right FCC by flight. The master FCC is not affected by the position of the Flight Director switches. The FCCs receive inputs from several systems including the air data inertial reference system (ADIRS). The AOA inputs are provided to each FCC by associated air data inertial reference unit (ADIRU). Each ADIRU receives AOA information from one of the two resolvers contained within the associated AOA sensor (i.e. the Left ADIRU uses left AOA vane and the Right ADIRU uses the right AOA vane).

The MCAS software uses input from a single AOA sensor only. Certain failure or anomalies of the AOA sensor corresponding to the master FCC controlling STS can generate an unintended activation of MCAS. Anticipated flight crew response including aircraft nose up (ANU) electric trim commands (which reset MCAS) may cause the flight crew difficultly in controlling the aircraft.

Erroneous AOA signals are not frequent events. Boeing reported that 25 activations of stick shaker mostly due to AOA failures occurred in 737 aircraft for the past 17 years during more than 240 million flight hours. The MCAS architecture with redundant AOA inputs for MCAS could have been considered but was not required based on the FHA classification of Major.

The accident flight was the first flight of the day therefore after powering up the aircraft, the left FCC acted as the operating STS channel and received input from left AOA sensor. The DFDR showed that on the accident flight, the left AOA sensor had a high bias 21° when compared to the right AOA sensor.

The FCC received the biased AOA data and, after the flaps were retracted, commanded MCAS. As stated earlier, a combination of repetitive MCAS-commands coupled with insufficient flight crew electric trim inputs and not using the STAB TRIM CUTOUT switches can lead to significant miss-trim and loss of control.

As discussed earlier, the condition remained hazardous since the training or the guidance for action taken in such situation were not adequately provided. To comply with the safety requirement of a "hazardous failure condition," the aircraft is supposed to rely on sensors that have less than a one-in-10-million (1E-7) chance of failing. Generally, that means taking measurements from two sensors. A hazardous failure condition depending on a single sensor should have been avoided in the certification process.

During the certification process, unintended MCAS activation was considered to have a major effect in the normal flight envelope and hazardous effect in the operational flight envelope. This means that a failure condition with a probability of occurrence less than or equal one-in-a hundred thousand (1E-5) chance would suffice in normal flight envelope. At the time of the accident roughly 250 737 MAX aircraft were in service with assumed average utilization of 2,000 flight hours, with the total accumulated fleet at 500,000 flight hours. With two AOA sensors installed on each aircraft, this means a failure occurred on the 737 MAX every 1,000,000 flight hours (1E-6), more than the requirement probability of 1E-7. The same AOA sensor is used on other aircraft and the AOA failure rate is even lower.

If the probability of an undesirable failure condition is not below the maximum allowable probability for that category of hazard, redesign of the system should be considered. If the uncommanded MCAS failure condition had been assessed as more severe than Major, the decision to rely on single AOA sensor should have been avoided.

MCAS Function

The MCAS specifications called for the system to limit its ability to move the horizontal stabilizer at the flaps up high rate trim up to a maximum of 2.5°. The DFDR recorded the changes of horizontal stabilizer as changes in pitch trim unit. In the accident flight, MCAS repeatedly moved the horizontal stabilizer based on a combination of the erroneous AOA inputs and flight crew manual electric inputs. The Captain managed to control the aircraft with the pitch trim. The DFDR data showed that control was maintained by keeping pitch trim above 5 units to counter the repetitive MCAS activations. After transfer of control to the FO, the FO did not apply sufficient manual electric trim to counter repetitive MCAS activations resulting in compounding mis-trim which required significant control column force. The DFDR data indicated the aircraft descended and could not be controlled. Column forces exceeded 100 lbs., which is more than the 75 lbs. limit set by the regulation (14 FAR 25.143).

The early design of MCAS allowed a maximum stabilizer movement of 0.6°, but it was later revised and extended to 2.5°. Boeing reasoned that unintended MCAS activation due to erroneous AOA input was able to be managed by using elevator alone, electric stabilizer trim or by stabilizer trim cutout switches. During manual flight, stabilizer manual electric trim is normally used to offload control column forces during mis-trim conditions. Without the awareness of the MCAS function, the flight crew would possibly recognize an MCAS activation as Speed Trim System (STS) input. MCAS behaves differently than the STS, it moves the horizontal stabilizer at a faster rate.

Following an erroneous MCAS activation, insufficient manual electric trim inputs result in the stabilizer not fully returning to its original trimmed position and gradually moving the stabilizer to greater mis-trim.

Any out of trim condition which is not properly corrected would lead the flight crew into a situation that makes it more difficult for them to maintain desired attitude of the aircraft. The flight crews in both the accident flight and the previous flight had difficulty maintaining flight path during repetitive MCAS activations.

Per Boeing, stabilizer trim cutouts switches were available but not required to counter MCAS activations. The only procedure that directs selecting the stabilizer cutout switches is the Runaway Stabilizer non-normal checklist (NNC). This NNS is used to stop un-commanded stabilizer trim wheel movement, which would stop MCAS-commanded stabilizer trim movement.

However, erroneous MCAS activation does not look like a typical stabilizer runaway, which is continuous un-commanded (runaway) movement of the stabilizer. During the accident flight, the stabilizer movement was not continuous; the MCAS commands were bounded by the MCAS authority (up to 2.5°); the pilots were able to counter the nose-down movement using opposing manual electric trim inputs; and after the pilots released the manual electric input and MCAS was reset, there was not another MCAS command for 5 seconds.

To incorporate MCAS, the basic column cutout function had to be inhibited during the MCAS activation. Pulling back on the column normally interrupts any electric stabilizer aircraft nose-down command, but for the MAX with MCAS operating, that control column cutout function is disabled.

Movement of the stabilizer due to uncommanded MCAS activation during normal flight would be easier to identify if there were no other distractions in the cockpit. However, during the accident flight erroneous inputs, as a result of the misaligned resolvers, from the AOA resulted in several fault messages (IAS DISAGREE, ALT DISAGREE on the PFDs, and Feel Differential Pressure light) that affected the flight crew's understanding and awareness of the situation. The stick shaker activated continuously after lift-off and the noise could have interfered with the flight crew hearing the sound of the stabilizer trim wheel spinning during MCAS operations. Therefore, the movement of stabilizer wheel might not have been recognized by the flight crew.

The flight crew of LNI043 eventually observed and recognized the un-commanded stabilizer movement and moved the stabilizer trim cutout switches to the cutout position. Stopping the stabilizer movement enabled the flight crew to continue the flight using manual trim wheel to control stabilizer position. On that flight, stabilizer cutout was used to counter the repetitive MCAS-commanded stabilizer.

The aircraft design should not have allowed this situation. The flight crew should have been provided with information and alerts to help them understand the system and know how to resolve potential issues. Flight crew procedures and training should be appropriate. The aircraft should have included the intended AOA DISAGREE alert message functionally, which was installed on 737 NG aircraft. Boeing and the FAA should ensure that new and changed aircraft design are properly described, analyzed, and certified.

Revision to MCAS Authority
The original MCAS design limited its authority to 0.6° of stabilizer movement, out of a physical maximum of 4.2° of nose-down movement. Later in the development program, the limit was increased to 2.5° after flight tests showed that additional movement of the stabilizer was required during elevated AOA conditions at low speeds.

The behavior of an aircraft in a high angle-of-attack stall is difficult to model in advance purely by analysis and during flight tests on a new aircraft, it is not uncommon to modify the control software to refine the aircraft's flight characteristics.

The higher limit meant that each time MCAS was triggered in a low Mach environment, it caused a much greater movement of the stabilizer than was specified in that original safety analysis document. As MCAS pushed the nose down, the flight crew could respond by operating the electric stabilizer trim switch long enough to reset the stabilizer movement. However, if the reset was not fully completed, subsequent activation of MCAS would move the aircraft further nose-down. At a limit of 2.5°, two cycles of MCAS without correction would have been enough to reach the maximum nose-down condition. At this point, the flight crew would experience extreme difficulty to maintain control of the aircraft.

After the change, FHA was reviewed but not all documents including Stabilizer System Safety Assessment were updated. Boeing briefed the FAA flight test prior to the certification flight testing that the MCAS envelope (low speed MCAS with the greater 2.5° authority) was expanding and it also updated the MCAS description in the EDFCS System Description document and provided it to the FAA.

Without documenting the updated analysis in the stabilizer SSA document, the FAA flight control systems specialists may not have been aware of the design change. Boeing did not submit the required documentation and the FAA did not sufficiently oversee Boeing ODA.

Decision to not Include Description in FCOM or Differences Training
Boeing and the FAA engaged in extensive discussion about the appropriate content of Boeing 737-8 (MAX) training and manuals for a period of several years prior to Boeing 737-8 (MAX) certification. During discussions and communications with the FAA beginning in March 2016, Boeing proposed removing MCAS from the FCOM and differences

tables. The supporting rationale was discussed between Boeing and FAA and accepted by FAA, but not formally documented in meeting minutes. Specifically, 14 FAR 25.1585 (b) states that *"Information or procedures not directly related to airworthiness or not under the control of the crew must not be included, nor must any procedure that is accepted as basic airmanship."*

Boeing considered that MCAS function is automatic, without any control input from the flight crew and that it would operate in the background. Boeing also considered that the procedure required to respond to any MCAS function was no different than the existing procedures and that crews were not expected to encounter MCAS in normal operation. The discussions did not consider the failure scenario seen on the accident flight.

The KNKT believes that the effect of erroneous MCAS function was startling to the flight crews. On 5 July 2017, the FAA issued the final FSB Report, which did not include MCAS in the differences table. Throughout this period, Boeing communicated about these issues with personnel in the FAA Airplane Evaluation Group, which chaired the FSB.

The lack of MCAS information in the FCOM and flight crew training resulted in it being more difficult for the flight crew to diagnose the problem and find the corrective procedure to solve it. Without prior awareness of the MCAS function, it would be more difficult for the flight crew to understand the problem. It would take them longer time to understand the situation and come to the correct solution, putting them at a higher risk than necessary. The flight crew of the previous LNI043 flight took around 3 minutes and 40 seconds after 12 MCAS activations to come up to the solution of the problem by performing stabilizer trim cut-out, while in the accident flight the flight crew did not manage to find the solution. Therefore, KNKT believes that flight crew should have been made aware of MCAS which would have provided them with awareness of the system and increase their chances of being able to mitigate the consequences of repetitive activations in the accident scenario.

Even though the flight crew can provide a nose-up input on the control column and/or manual electric trim every time the MCAS activates, the MCAS software continues to work against them, making it more difficult to fully control the aircraft if column forces resulting from MCAS inputs are not completely negated by the flight crew. This condition meant that the flight crew were running out of time to find solution before the repetitive MCAS activations without fully retrimming the aircraft placed the aircraft the flight into in an extreme nose-down attitude that the flight crew was unable to recover from.

The unintended MCAS-commanded stabilizer movement due to erroneous AOA input was considered a failure condition with Major effect in normal flight envelope.

Flight crew training would have supported the recognition of abnormal situations and appropriate flight crew action. Boeing did not provide additional training requirements for the Boeing 737-8 (MAX) since the condition was considered similar to previous 737 models.

As stated earlier, Boeing assumed that it was appropriate to rely on trained crew procedures, especially if these procedures are considered memory items. However, since the procedures are not included in the transition training, and any means for identifying flight situations arising from the difference in design is not adequately provided, this assumption is invalid.

Summary

The MCAS function was not a fail-safe design and did not include redundancy. A single failure to the AOA sensor corresponding with the FCC commanding STS resulted in erroneous activation of MCAS. During the accident, flight crew reactions were different from and did not match the guidance for assumptions of flight crew behavior that were used when classifying the hazard severity of this failure mode in the functional hazard assessment.

Classification of unintended MCAS-command as Major in the FHA resulted in a system design that was inadequate for this failure.

Not including information about the MCAS from the FCOM and flight crew training made it more difficult for the flight crew to diagnose problems and find the corrective actions to overcome the situation.

The absence of an AOA Disagree message made it more difficult for the flight crew to diagnose the failure and for maintenance to diagnose and correct the failure.

Regulations

The 14 FAR 25 has outlined the regulations related to the safety and airworthiness of large transport category aircraft. Revisions have been made to make sure the sufficiency of regulations to ensure safety. For the safety assessment of aircraft systems, the 14 FAR 25.1309 set the requirements for the design and installation of systems which include analysis of effects and probabilities of single, multiple and combined failures of systems. It assumed that flight crew would correctly respond to flight conditions in case of such failures. Human error is not included in the probability analysis, even though the flight crew is often used as a means to mitigate a failure condition.

The accident flights, system malfunction led to erroneous information that initiated a series of events that were not correctly recognized and responded by the flight crew. Therefore, KNKT recommends FAA to work with international regulatory authorities to review assumptions on flight crew behavior used during design and revise certification processes to ensure assumptions used during the design process are validated.

When performing safety assessments to comply with 14 FAR 25.1309, Boeing followed the procedures set in FAA AC 25.1309-1A and the SAE ARP 4761 as the acceptable means of compliance. When doing the analysis, Boeing assumed that the flight crew are completely reliable and would respond correctly and appropriately to the situations in time. During the accident and previous LNI043 flights, some of these assumptions were incorrect, since the flight crew responded differently from what was expected.

14 FAR 25.671 (c) requires that probable malfunctions of the flight control system must be capable of being readily counteracted by the flight crew. This necessitates that normal flight crew should be able to readily identify problems and respond quickly to mitigate them. However, during the accident flight multiple alerts and indications concealed the actual problem and made it difficult for the flight crew to understand and mitigate it.

14 FAR 25.672 further requires that the design of the stability augmentation system or of any other automatic or power-operated system such as MCAS must permit initial counteraction of failures without requiring exceptional flight crew skill or strength. This means that an average flight crew should be able to override or mitigate the failure by movement of the flight controls in the normal sense. The accident flight and previous flight provided evidence that mitigating the repetitive erroneous activations of MCAS required skills that were different than what was expected. Pilots have been expected to recognize an out of trim condition and correct it by re-trimming.

During certification phase, compliance was demonstrated by flight test pilots which normally have exceptional skill and experience. Flight test pilots generally have more knowledge about the aircraft design characteristics than normal pilots. This level of competence usually cannot be translated to most pilots however, test pilots are trained to replicate the average flight crew. The Aircraft Evaluation Group pilots, who have an operational flying background, also evaluate the aircraft during the certification phase. These pilots establish the flight crew type rating, training, checking and currency requirements as part of the Flight Standardization Board (FSB) process. The FSB process for Boeing 737-8 (MAX) utilized airline line pilots to help ensure the requirements are operationally representative. The FAA and OEMs should re-evaluate their assumptions for what constitutes an average flight crew's basic skill and what level of systems knowledge a 'properly trained average flight crew' has when encountering failures.

FAA Oversight

In 2009, the FAA transitioned to a delegation process for Boeing, over seen by FAA's Boeing Aviation Safety Oversite Office (BASOO). This delegation of certification functions is done through a formal procedure, under the system of Organization Designation Authorization (ODA). Boeing ODA is authorized to select and appoint individuals to perform some of the delegated functions as representatives of FAA. The delegated functions for a Type Certification (TC) ODA are:

- establishing and determining conformity of parts, assemblies, installations, test setups, and products (aircraft);
- finding compliance with airworthiness standards for new design, or major changes to design;
- issuing special flight permits for operation of aircraft;
- issuing -airworthiness approvals for articles (Export), and aircraft (Standard or Export)

The BASOO conducts routine audits/supervisions, large-scale inspections and post-project reviews and was involved in safety critical and novel and unusual areas such as MCAS in the review and acceptance of certification plans.

The MCAS system design was based on an FHA classification of Major for uncommanded MCAS activation, which was made using FAA-published guidance material regarding assumptions of flight crew response. Those assumptions were:

- Uncommanded system inputs are readily recognizable and can be counteracted by overriding the failure by movement of the flight controls in the normal sense by the flight crew and do not require specific procedures.

- Action to counter the failure shall not require exceptional piloting skill or strength.
- The flight crew will take immediate action to reduce or eliminate increased control forces by re-trimming or changing configuration or flight conditions.
- Trained flight crew memory procedures shall be followed to address and eliminate or mitigate the failure.

At this time these assumptions are industry standards used by both the FAA and Boeing. The assumptions are what are used in the classification of the hazard.

In the accident flight, the system malfunction led to a series of aircraft and flight crew interactions which the flight crew did not understand or know how to resolve. It is the flight crew response assumptions in the initial design process which, coupled with the repetitive MCAS activations, turned out to be incorrect and inconsistent with the FHA classification of Major.

Xtra Aerospace LLC

The Xtra Aerospace utilized the Angle Position Indicator (API) Peak Model SRI-201B (Model 7724-00-2) for test and calibration repair of the accident AOA sensor, part number 0861FL1 serial number 14488. Utilization of the Peak API deviates from the equipment described in CMM Revision 8, and thus required an equivalency assessment by the Xtra Aerospace engineering department and subsequent acceptance by local FAA Flight Standards District Office (FSDO).

Representatives of NTSB, FAA, Boeing and Collins Aerospace reviewed the test equipment equivalency report and observed that according to the report the Peak SRI-201B was capable of performing all tests required by the OEM, of checking all required parameters and met the level of accuracy specified by the CMM. A subsequent comparison of the accuracy specifications found that the Peak SRI-201B API accuracy (+/- .03°) was not equivalent to the CMM recommended North Atlantic 8810 (+/- .004°). Although Note - the CMM revision 8 most restrictive requirement is +/- 0.03°for the resolver vane 0° position. Nor did the investigation find a written instruction to operate the Peak Electronics SRI-201B API.

Despite the lack of API specific written instructions for the alternate equipment, Xtra Aerospace nevertheless obtained acceptance of their API equipment equivalency report from the FAA FSDO.

The participants involved in the visit concluded that performing the required testing and calibration defined in CMM Revision 8 using the Peak API could potentially introduce a bias into both resolvers if the REL/ABS (Relative/Absolute) switch on the Peak Electronics API was inadvertently positioned to REL. Repeated tests conducted at Collins Aerospace in February 2019 and at Boeing in June 2019 supported the same conclusion.

Xtra Aerospace records reflect that repair activities of AOA sensor S/N 14488 included replacement of the AOA vane-slinger-shaft assembly (VSS). During VSS removal and replacement, it is highly unlikely that the resolvers retained their original position. During resolver calibration, utilization of the Peak API and selection of the REL/ABS switch in the REL position selection may have led to improper calibration because there was no written instruction for correct utilization of the Peak API in accordance with the CMM requirements. With the REL/ABS switch in the REL position for CMM return-to-service testing, the improper calibration would not be detected.

After Xtra Aerospace repair of the accident AOA sensor in November of 2017, the sensor was installed on the PK-LQP aircraft on left side position during the maintenance activity in Denpasar on 28 October 2018. On the subsequent flight, a 21° difference between left and right AOA sensors was recorded on the DFDR, commencing shortly after the takeoff roll was initiated (note it takes some airflow over the vane for the vanes to align with the airflow). This immediate 21° delta indicated that the AOA sensor was most likely improperly calibrated at Xtra Aerospace.

As noted, utilization of the Peak Model SRI-201B API by Xtra Aerospace for the test and calibration of the 0861FL1 AOA sensor should have required a written procedure to specify the proper position of the REL/ABS switch.

The lack of an API written procedure was not detected by the FAA's FSDO. This indicates inadequacy of FAA oversight.

CHAPTER 8

CONCLUSIONS

Findings

Findings are statements of all significant conditions, events or circumstances in the accident sequence. The findings are significant steps in the accident sequence, but they are not always causal, or indicate deficiencies. Some findings point out the conditions that pre-existed the accident sequence, but they are usually essential to the understanding of the occurrence, usually in chronological order.

The KNKT identified findings as follows:

1. MCAS is designed to function only during manual flight (autopilot not engaged), with the aircraft's flaps up, at an elevated AOA. As the development of the 737-8 (MAX) progressed, the MCAS function was expanded to low Mach numbers and increased to maximum MCAS command limit of 2.5□ of stabilizer movement.

2. During the Functional Hazard Analysis (FHA), unintended MCAS-commanded stabilizer movement was considered a failure condition with Major effect in the normal flight envelope. The assessment of Major did not require Boeing to more rigorously analyze the failure condition in the safety analysis using Failure Modes and Effects Analysis (FMEA) and Fault Tree Analysis (FTA), as these are only required for Hazardous or Catastrophic failure conditions.

3. Uncommanded MCAS function was considered Major during the FHA. Boeing reasoned that such a failure could be countered by using elevator alone. In addition, stabilizer trim is available to offload column forces, and stabilizer cutout is also available but not required to counter failure.

4. FMEA would have been able to identify single-point and latent failures which have significant effects as in the case of MCAS design. It also provides significant insight into means for detecting identified failures, flight crew impact on resolution of failure effect, maintenance impact on isolation of failure and corresponding restitution of system.

5. Boeing conducted the FHA assessment based on the FAA guidance and was also based on an assumption that the flight crew was highly reliable to respond correctly and in time within 3 seconds. The assessment was that each MCAS input could be controlled with control column alone and subsequently re-trimmed to zero column force while maintaining flight path.

6. The flight crew did not react to MCAS activation but to the increasing force on the control column. Since the flight crew initially countered the MCAS command using control column, the longer response time for making electric stabilizer trim inputs was understandable.

7. During the accident and previous LNI043 flights, the flight crew initially responded by pulling back on the control column, however, they did not consistently trim out the resulting column forces as had been assumed. As a result the Boeing assumption was different from the flight crew behavior in responding to MCAS activation.

8. During FHA, the simulator test had never considered a scenario in which the MCAS activation allowed the stabilizer movement to reach the maximum MCAS limit of 2.5 degrees. Repetitive MCAS activations without adequate trim reaction by the flight crew would make the stabilizer move to maximum deflection and escalate the flight crew workload and hence failure effects should have been reconsidered. Therefore, their combined flight deck effects were not evaluated.

9. In the event of multiple MCAS activations with repeated electric trim inputs by flight crew without sufficient response to return the aircraft to a trimmed state, the control column force to maintain level flight could eventually increase to a level where control forces alone may not be adequate to control the aircraft. The cumulative mis-trim could not be countered by using elevator alone which is contrary to the Boeing assumption during FHA.

10. Any out of trim condition which is not properly corrected would lead the flight crew into a situation that makes it more difficult for them to maintain desired attitude of the aircraft. The flight crews in both the accident flight and the previous flight had difficulty maintaining flight path during multiple MCAS activations.

11. The procedure of runaway stabilizer was not reintroduced during transition training and there was no immediate indication available to the flight crew to be able to directly correlate the uncommanded nose down stabilizer to the procedure. Therefore, the assumption of relying on trained crew procedures to implement memory items was inappropriate

12. During the accident flight, multiple alerts and indications occurred which increased flight crew's workload. This obscured the problem and the flight crew could not arrive at a solution during the initial or subsequent automatic aircraft nosedown stabilizer trim inputs, such as performing the runaway stabilizer procedure or continuing to use electric trim to reduce column forces and maintain level flight.

13. In the event of MCAS activation with manual electric trim inputs by the flight crew, the MCAS function will reset which can lead to subsequent MCAS activations. To recover, the flight crew has 3 options to respond, if one of these 3 responses is not used, it may result in a miss-trimmed condition that cannot be controlled.

14. The flight crew of LNI043 eventually observed and recognized the uncommanded stabilizer movement and moved the stabilizer trim cutout switches to the cutout position. Stopping the stabilizer movement enabled the flight crew to continue the flight using manual trim wheel to control stabilizer position. On that flight, stabilizer cutout was used to counter the repetitive MCAS-commanded stabilizer. Boeing reasoning that the stabilizer cutout is available but not required is incorrect.

15. Boeing considered that the loss of one AOA and erroneous AOA as two independent events with distinct probabilities. The combined failure event probability was assessed as beyond extremely improbable, hence complying with the safety requirements for the Air Data System. However, the design of MCAS relying on input from a single AOA sensor, made this Flight Control System susceptible to a single failure of AOA malfunction.

16. During the single and multiple failure analysis from the air data system worst case scenario of "failure of one AOA followed by erroneous AOA", Boeing concluded that the effect would be hazardous until the flight crew

recognized the problem and took appropriate action to mitigate it. Since the training or the guidance for actions taken in such situation were not provided, the effect category should have remained hazardous.

17. Since the FCC controlling the MCAS is dependent on a single AOA source, the MCAS contribution to cumulative AOA effects should have been assessed.

18. The MCAS software uses input from a single AOA sensor only. Certain failures or anomalies of the AOA sensor corresponding to the master FCC controlling STS can generate an unintended activation of MCAS. Anticipated flight crew response including aircraft nose up (ANU) electric trim commands (which reset MCAS) may cause the flight crew difficultly in controlling the aircraft.

19. The MCAS architecture with redundant AOA inputs for MCAS could have been considered but was not required based on the FHA classification of Major.

20. If the probability of an undesirable failure condition is not below the maximum allowable probability for that category of hazard, redesign of the system should be considered. If the uncommanded MCAS failure condition had been assessed as more severe than Major, the decision to rely on single AOA sensor should have been avoided.

21. The DFDR data indicated that during the last phase of the flight, the aircraft descended and could not be controlled. Column forces exceeded 100 pounds, which is more than the 75-pound limit set by the regulation (14 CFR 25.143).

22. Pulling back on the column normally interrupts any electric stabilizer aircraft nose-down command, but for the 737-8 (MAX) with MCAS operating, that control column cutout function is disabled.

23. During the accident flight erroneous inputs, as a result of the misaligned resolvers, from the AOA sensor resulted in several fault messages (IAS DISAGREE, ALT DISAGREE on the PFDs, and Feel Differential Pressure light) and activation of MCAS that affected the flight crew's understanding and awareness of the situation.

24. The stick shaker activated continuously after lift-off and the noise could have interfered with the flight crew hearing the sound of the stabilizer trim wheel spinning during MCAS operations. Therefore, the

movement of stabilizer wheel might not have been recognized by the flight crew

25. The aircraft design should provide the flight crew with information and alerts to help them understand the system and know how to resolve potential issues.

26. Boeing did not submit the required documentation and the FAA did not sufficiently oversee Boeing ODA. Without documenting the updated analysis in the stabilizer SSA document, the FAA flight control systems specialists may not have been aware of the design change.

27. Boeing considered that MCAS function is automatic, the procedure required to respond to any MCAS function was no different than the existing procedures and that crews were not expected to encounter MCAS in normal operation therefor Boeing did not consider the failure scenario seen on the accident flight. The investigation believes that the effect of erroneous MCAS function was startling to the flight crews.

28. The investigation believes that flight crew should have been made aware of MCAS which would have provided them with awareness of the system and increase their chances of being able to mitigate the consequences of multiple activations in the accident scenario.

29. Without understanding of MCAS and reactivation after release the electric trim, the flight crew was running out of time to find a solution before the repetitive MCAS activations without fully retrimming the aircraft placed the aircraft into in an extreme nose-down attitude that the flight crew was unable to recover from.

30. Flight crew training would have supported the recognition of abnormal situations and appropriate flight crew action. Boeing did not provide information and additional training requirements for the 737-8 (MAX) since the condition was considered similar to previous 737 models.

31. The aircraft should have included the intended AOA DISAGREE alert message functionally, which was installed on 737 NG aircraft. Boeing and the FAA should ensure that new and changed aircraft design are properly described, analyzed, and certified.

32. The absence of an AOA Disagree message made it more difficult for the flight crew to diagnose the failure and for maintenance to diagnose and correct the failure.

33. For the safety assessment of aircraft systems, the 14 FAR 25.1309 set the requirements for the design and installation of systems which include analysis of effects and probabilities of single, multiple and combined failures of systems. It assumed that flight crew would correctly respond to flight conditions in case of such failures. Human error is not included in the probability analysis, even though the flight crew is often used as a means to mitigate a failure condition.

34. When performing safety assessments to comply with 14 FAR 25.1309, Boeing followed the procedures set in FAA AC 25.1309-1A and the SAE ARP 4761 as the acceptable means of compliance. When doing the analysis, Boeing assumed that the flight crew are completely reliable and would respond correctly and appropriately to the situations in time. During the accident and previous LNI043 flights, some of these assumptions were incorrect, since the flight crew responded differently from what was expected.

35. 14 FAR 25.671 (c) requires that probable malfunctions of the flight control system must be capable of being readily counteracted by the flight crew. This necessitates that normal flight crew should be able to readily identify problems and respond quickly to mitigate them. However, during the accident flight multiple alerts and indications concealed the actual problem and made it difficult for the flight crew to understand and mitigate it.

36. The Flight Standardization Board (FSB) process for the Boeing 737-8 (MAX) utilized airline line pilots to help ensure the requirements are operationally representative. The FAA and OEMs should re-evaluate their assumptions for what constitutes an average flight crew's basic skill and what level of systems knowledge a 'properly trained average pilot' has when encountering failures.

37. In the accident flight, the system malfunction led to a series of aircraft and flight crew interactions which the flight crew did not understand or know how to resolve. It is the flight crew response assumptions in the initial design process which, coupled with the repetitive MCAS activations, turned out to be incorrect and inconsistent with the FHA classification of Major.

38. The first problem reported on PK-LQP aircraft of SPD and ALT flags appeared on Captain's PFD occurred on 26 October 2018 during the flight from Tianjin to Manado and reappeared 3 times within 5 flight sectors.

39. The SPD and ALT flag did not occur on the flight from Denpasar to Lombok and return. This was consistent with the result of AOA sensor examination which indicated that the resolver 2 became unreliable during cold temperature.

40. The engineer in Manado suggested to the flight crew to continue the flight as problem rectification would be better to be performed in Denpasar and considering that the SPD and ALT flags had no longer appeared on the Captain's PFD. This indicated that the aircraft was released with known possible recurring problem.

41. On the flight from Manado to Denpasar on 28 October 2018, the DFDR recorded the A/T disengaged on takeoff roll and the SPD and ALT flags on the captain's PFD most likely had appeared after the engine start. The altimeter and speed indicator are airworthiness related instruments and must be serviceable for dispatch. The decision to continue the flight was contrary to the company procedure.

42. The engineer in Denpasar considered that the problem had appeared repeatedly and decided to replace the left AOA sensor. Replacement of AOA sensor proved to be the solution to rectify the SPD and ALT flags that were reported to have appeared on the Captain's PFD, however the installed AOA sensor was misaligned by about 21° and resulted in different problems.

43. The Boeing test result indicated that a misaligned AOA sensor would not pass the installation test as the AOA values shown on the SMYD computer were out of tolerance and "AOA SENSR INVALID" message appeared in the SMYD BITE module. This test and subsequent testing verified that the alternate method of the installation test could identify a 20 or 21° bias in the AOA sensor.

44. Comparing the results of the installation test in Denpasar and Boeing, the investigation could not determine that the AOA sensor installation test conducted in Denpasar with any certainty.

45. The BAT LMPM required the engineer to record the test values to ensure that the test results were within tolerance. The engineer did not record the value of the AOA angle deflection during the AOA sensor installation test. Therefore, neither BAT nor Lion Air identified that the documentation had not been filled out.

46. After LNI043 was airborne, the left control column stick shaker was active and several messages appeared. The Captain of LNI043 was aware

to the aircraft condition after discussion with the engineer in Denpasar. This awareness helped the Captain to make proper problem identification.

47. The Captain action of transferring the control prior to crosscheck of the instruments may have indicated that the Captain generally was aware of the repetitive previous problem of SPD and ALT flags and the replacement of the left AOA sensor on this aircraft.

48. The LNI043 flight crew performed NNC of Runaway Stabilizer Trim by selecting the STAB TRIM switches to cut-out, which resulted in termination of AND activations by MCAS, and the aircraft became under control with consequences of inability to engage the autopilot, and requirement for manual operation of stabilizer trim by hand.

49. The Captain's decision to continue to the destination was based on the fact that a requirement to "land-at-the-nearest-suitable-airport" in the three Non-Normal-Checklists was absent.

50. The Captain of LNI043 felt confident to continue the flight to the destination because the aircraft was controllable and the expected weather along the route and at the destination was good.

51. The LNI043 flight crew decision to continue with stick shaker active is not common in comparison to previous events of erroneous stick shaker. When combined with the runaway stabilizer situation recognized by the flight crew, the decision to continue was highly unusual.

52. During the descent to destination they requested uninterrupted descent path profile. This action suggested that the flight crew were aware of their existing flight condition (continuous stick shaker, manual flying, manual trimming, FO PFD was the primary instrument) required a simplified flight path management until approach and landing.

53. During flight, the Captain of LNI043 kept the fasten seat belt sign on and asked the deadheading flight crew to assist the cockpit tasks. These actions indicated that the Captain was aware of the need to use all available resources to alleviate the matter to complete the flight to the destination, despite the increased workload and stressful situation.

54. The AFML entry for LNI043, which did not contain additional details about what was experienced, was not in accordance with company guidance provided in OM-Part A, Section 11.4.9 which lists reportable events to include "Warning or alert, including flight control warnings, door warnings, stall warning (stick-shaker), fire/smoke/fumes warning."

55. The SS Directorate did not notice the occurrence since the report was filed outside normal office hours and the report to SS Directorate was not processed until the office hours on the following day.

56. The insufficient SMS training and inability of the employees to identify the hazard might also be indicated by the incomplete post-flight report of the problems that occurred on LNI043. The incomplete report became a hazard as the known or suspected defects were not reported which might make the engineer unable to properly maintain the airworthiness of the aircraft.

57. Content of the report did not trigger the Duty Management Pilot to assess this as a Serious Incident and enable a safety investigation. The risk of the problems that occurred on the flight LNI043 were not assessed to be considered as a hazard on the subsequent flight.

58. The LNI043 flight that experienced multiple malfunctions were considered caused or could have caused difficulties in controlling the aircraft. According to the ICAO Annex 13, CASR part 830 and OM-part A, the flight is classified as serious incident which required investigation by the KNKT in accordance with the Aviation Law Number 1 of 2009 and Government Decree Number 62 of 2013.

59. The definition of an aircraft repetitive problem was different between Lion Air CMM and BAT AMOQSM. This difference indicated that the Lion Air did not monitor the repetitive problem policy of the BAT as a subcontracted entity.

60. The requirement to report all known and suspected defects is very critical for engineering to be able to maintain the airworthiness of the aircraft.

61. The fault code was not documented in the AFML. The engineer did not record the maintenance message that appeared in the OMF in the AFML. Being unaware of the maintenance message and the fault code, this would increase the difficulty for trouble shooting by the engineer.

62. The IFIM tasks of "ALT DISAGREE" and "IAS DISAGREE" have repetition on the leak test in steps (3) and (4) as they are referring to the same AMM tasks. This repetition was inefficiency and does not contribute to the problem solving.

63. The inhibited AOA DISAGREE message contributed to the inability of the engineer to rectify the problems that occurred on the LNI043 flight which were caused by AOA sensor bias.

64. The lack of an AOA DISAGREE message did not match the Boeing system description that was the basis for certifying the aircraft design. The software not having the intended functionality was not detected by Boeing nor the FAA during development and certification of the 737-8 MAX before the aircraft had entered service.

65. During the LNI610 flight preparation, the CVR did not record flight crew discussion about previous aircraft problems recorded in the AFML. This might have made the flight crew of LNI610 would not be aware of aircraft problems that might reappear during flight, including the stick shaker activation and uncommanded AND trim. This would lead to the inability of the flight crew to predict and be prepared to mitigate the events that might occur.

66. Just after liftoff, the left stick shaker activated and numerous messages on the PFD were displayed, repetitive MCAS activation after the flaps were retracted and the ATCo communication increased the flight crew workload.

67. The FO asked the controller of the aircraft altitude and the indicated speed on the ATC radar display in an attempt to obtain another source of information. However, the ATC radar receives altitude data transmitted by the aircraft therefore, no additional data may be acquired. Being unable to determine reliable altitude and airspeed might increase stress to the flight crew.

68. The inability for the FO to perform memory items and locate the checklist in the QRH in a timely manner indicated that the FO was not familiar with the NNC. This condition was reappearance of misidentifying NNC which showed on the FO's training records.

69. Despite the flight crew's attempt to execute the NNC, due to increased workload, and distractions from the ATC communication, the NNC was unable to be completed in that situation. The unfinished NNC made it difficult for the flight crew of LNI610 to understand the aircraft problem and how to mitigate the problem.

70. The reappearance of difficulty in aircraft handling identified during training in the accident flight indicated that the Lion Air training rehearsal was not effective.

71. The controller provided eight heading instructions after the flight crew reported that the aircraft was experiencing a flight control problem, which was not considered as an emergency condition according to ATS SOP of AirNav Indonesia branch JATSC. There was also no objection by the flight crew to the heading instructions and the flight crew did not declare an emergency. These conditions increased the flight crew workload.

72. The absence of a declaration of urgency (PAN PAN) or emergency (MAYDAY), or asking for special handling, resulted in the ATCo not prioritizing that flight. With priority, ATC would not require LNI610 to maneuver repeatedly.

73. The AOA DISAGREE message was inhibited on the accident aircraft therefore, flight crews would not be aware that this message would not appear if the AOA DISAGREE conditions were met. This would contribute to flight crew being denied valid information about abnormal conditions being faced and lead to a significant reduction in situational awareness by the flight crew.

74. No information about MCAS was given in the flight crew manuals and MCAS was not included in the flight crew training. These made the flight crew unaware of the MCAS system and its effects. There were no procedures for mitigation in response to erroneous AOA.

75. Both flight crew of LNI610 being preoccupied with individual tasks indicated that the crew coordination was not well performed. The Captain and FO did not have a shared mental model of the situation as exhibited by their lack of clear and effective communication. Most of the components of effective crew coordination were not achieved, resulting in failure to achieve the common goal of flying the aircraft safely.

76. During the multiple MCAS activations, the Captain managed to control the aircraft altitude. The Captain did not verbalize to the FO the difficulty in controlling the aircraft and the need for repeated aircraft nose up trim. The FO was preoccupied with completing the NNC and not monitoring the flight progress. Subsequently, the FO did not provide adequate electric trim to counter multiple MCAS activations.

77. The requirement to describe specific handling situation to the flight crew receiving the control was not required per Lion Air procedure or Indonesia requirement. The absence of Captain's specific description contributed to the FO's difficulty to understand the situation and may have contributed to his inability to mitigate the problem.

78. The content of the manual of Lion Air and BAT contain several inconsistencies, incompleteness, and unsynchronized procedures.

79. The investigation found that the engineers were prone to entering the problem symptom reported by the flight crew in the IFIM first instead of reviewing the OMF maintenance message. Conducting this method might lead the engineers into the inappropriate rectification task.

80. The investigation found that all AFML pages received by the investigation did not contain fault codes. The absence of the fault code reported by the flight crew may increase the workload of the engineer and prolong the rectification process.

81. The investigation considered that the amount of time to cover the hazard identification topic in the SMS training syllabus was insufficient. This may reduce the ability of employees to define and report a hazard.

82. A subsequent comparison of the accuracy specifications found that the Peak SRI-201B API accuracy met the requirement stated on the CMM revision 8. The investigation did not find a written instruction to operate the Peak Electronics SRI-201B API.

83. Despite the lack of API specific written instructions for the alternate equipment, Xtra Aerospace nevertheless obtained acceptance of their API equipment equivalency report from the FAA FSDO. The lack of an API written procedure was not detected by the FAA's FSDO. This indicates inadequacy of FAA oversight.

84. The Xtra Aerospace visit concluded that performing the required testing and calibration defined in CMM Revision 8 using the Peak API could potentially introduce a bias into both resolvers if the REL/ABS (Relative/Absolute) switch on the Peak Electronics API was inadvertently positioned to REL.

85. The OMF has the history page which contains record of the aircraft problems which can be utilized as a source for aircraft problem monitoring. The BAT has not utilized the OMF information as the source of aircraft problem monitoring.

86. On the subsequent flight, a 21□ difference between left and right AOA sensors was recorded on the DFDR, commencing shortly after the takeoff roll was initiated. This immediate 21□ delta indicated that the AOA sensor was most likely improperly calibrated at Xtra Aerospace.

87. As noted, utilization of the Peak Model SRI-201B API by Xtra Aerospace for the test and calibration of the 0861FL1 AOA sensor should have required a written procedure to specify the proper position of the REL/ABS switch.

88. The aircraft was equipped with an airframe-mounted low frequency underwater locator beacon (ULB) which operated at a frequency of 8.8 kHz. The beacon was mounted on the forward side of the nose pressure bulkhead. During the search phase, multiple surveys were conducted to detect a signal at 8.8 kHz, however no such signals were detected in the area where wreckage was recovered.

89. On 10 March 2019, an accident related to failure of an AOA sensor occurred involving a Boeing 737-8 (MAX) registered ET-AVJ operated by Ethiopian Airlines for scheduled passenger flight from Addis Ababa Bole International Airport (HAAB), Ethiopia to Jomo Kenyatta International Airport (HKJK), Kenya with flight number ET-302.

Contributing Factors

Contributing factors defines as actions, omissions, events, conditions, or a combination thereof, which, if eliminated, avoided or absent, would have reduced the probability of the accident or incident occurring, or mitigated the severity of the consequences of the accident or incident. The presentation is based on chronological order and not to show the degree of contribution.

1. During the design and certification of the Boeing 737-8 (MAX), assumptions were made about flight crew response to malfunctions which, even though consistent with current industry guidelines, turned out to be incorrect.

2. Based on the incorrect assumptions about flight crew response and an incomplete review of associated multiple flight deck effects, MCAS's reliance on a single sensor was deemed appropriate and met all certification requirements.

3. MCAS was designed to rely on a single AOA sensor, making it vulnerable to erroneous input from that sensor.

4. The absence of guidance on MCAS or more detailed use of trim in the flight manuals and in flight crew training, made it more difficult for flight crews to properly respond to uncommanded MCAS.

5. The AOA DISAGREE alert was not correctly enabled during Boeing 737-8 (MAX) development. As a result, it did not appear during flight with the mis-calibrated AOA sensor, could not be documented by the flight crew and was therefore not available to help maintenance identify the mis-calibrated AOA sensor.

6. The replacement AOA sensor that was installed on the accident aircraft had been mis-calibrated during an earlier repair. This mis-calibration was not detected during the repair.

7. The investigation could not determine that the installation test of the AOA sensor was performed properly. The mis-calibration was not detected.

8. Lack of documentation in the aircraft flight and maintenance log about the continuous stick shaker and use of the Runaway Stabilizer NNC meant that information was not available to the maintenance crew in Jakarta nor was it available to the accident crew, making it more difficult for each to take the appropriate actions.

9. The multiple alerts, repetitive MCAS activations, and distractions related to numerous ATC communications were not able to be effectively managed. This was caused by the difficulty of the situation and performance in manual handling, NNC execution, and flight crew communication, leading to ineffective CRM application and workload management. These performances had previously been identified during training and reappeared during the accident flight.

CHAPTER 9

SAFETY ACTION

At the time of issuing this Final Report, the KNKT had been informed of safety actions taken by several parties resulting from this accident.

Lion Air

On 29 October 2018, the Safety and Security Directorate issued safety reminder to all Boeing 737 pilots to review several procedures including memory items of airspeed unreliable and runaway stabilizer.

On 30 October 2018, issued information to all pilots which contained reminder to:

- Have a thoroughly understanding on Deferred Maintenance Item (DMI) for the aircraft to be use.
- Check any defect and the trouble shooting on Aircraft Maintenance Flight Log (AFML) from the previous flights.
- Be ready for any abnormal or emergency conditions by having Memory Items and maneuvers reviewed and have a good Cockpit Resource Management (CRM) to all counterparts.
- Write on the AFML for any malfunctions that happened during the flight. Brief the engineer on duty comprehensively about the malfunction happened in flight. Please refer to Fault Reporting Manual (FRM) provided in the aircraft.

- Send report to Safety and Security Directorate through all reporting methods that available as soon as practicable.

On 2 November 2018, the Safety and Security Directorate issued safety instruction:

For Operation Directorate:

- To instruct all Boeing 737 pilots to use the Fault Reporting Manual (FRM) in all their Aircraft Flight Maintenance Log (AFML) report. This measure shall be enforced by Operations, Training and Standard with immediate effect.
- To instruct all pilots to fill the AFML report with as much details as deem necessary to provide a full comprehensive description of the technical defect to the engineering team. This measure should be applied with immediate effect.
- To reinforce in the current simulator syllabus, the "Unreliable Airspeed" and "Stabilizer Runaway" maneuvers, with immediate effect to all fleets.
- To reinforce the role of Chief Pilot on Duty, in order to raise operational issue to IOCC/MCC should any significant notification has been received. This measure should be applied with immediate effect.
- To reinforce through Notice to Pilots, Ground Recurrent Training, and Simulator Sessions on Decision Making Process when the aircraft has declared and operating in abnormal (PAN-PAN) or emergency (MAYDAY-MAYDAY) condition.

For Maintenance Directorate:

- To ensure Batam Aero Technic (BAT) reinforce the role of technical specialist team as line maintenance support for more efficient troubleshooting process. This service should ensure that the "live" malfunctions are properly followed up until properly solved.
- To ensure Batam Aero Technic (BAT) through their TRAX system gives adequate alert on repetitive problem, even though reports for a malfunction may have been coded under different ATA references.
- To reinforce the MCC role in malfunction follow up and troubleshooting.

On 3 November 2018, the Chief Pilot issued Notice to Pilot which required all pilots to perform the following:

- Read and study the FRM (Fault Reporting Manual) and know how to utilize it. Any observed faults, status message, or cabin faults must be written down in the AFML, and ATA Number/Tittle of ECAM Shown (Fault) For A330. Should have any doubt, please contact the chief pilot or Quality Assurance Department via Mission Control (MC) – OM-part A 8.6.8.

- Do not hesitate to describe in details about the defect that has been encountered. This is a good practice especially for the engineers to do the troubleshooting and for the next crew that will fly the aircraft.

- Review the memory item routinely during the briefing, and if applicable, review the course of actions that should be taken if particular situations occur in any phase of flight.

On 5 November 2018, the Training Manager issued Training Notice to Pilot which required all instructor pilots to make additional training of airspeed unreliable and runaway stabilizer.

On 7 November 2018, the Fleet Manager issued Notice to Pilot which required all pilots to improve reporting events of IAS disagree, ALT disagree, SPEED fail, and ALT fail as a serious occurrence.

On 8 November 2018, the Safety and Security Directorate issued Safety Instruction to all pilots to follow Boeing Flight Crew Operations Manual Bulletin Number TBC-19 and Number MLI-15.

On 12 November 2018, the Safety and Security Directorate issued Notice to all station and operation managers of the Emergency Flowchart revision which included occurrence involving urgency and distress call events to be reported through Emergency Response Report flow.

On 15 November 2018, the Safety and Security Directorate issued Safety Instruction to Safety Corporate Director and Batam Aero Technic (BAT) Director to implement Directorate General of Civil Aviation Airworthiness Directive number 18-11-011-U.

The Lion Air also conducted several corrective actions to respond the KNKT safety recommendation written in the preliminary report as follows:

04.O-2018-35.2

According to the weight and balance sheet, on board the aircraft were two pilots, five flight attendants and 181 passengers consisted of 178 adult, one child and two infants. The voyage report showed that the number of flight attendant on board was six flight attendants. This indicated that the weight and balance sheet did not contain actual information.

KNKT recommend ensuring all the operation documents are properly filled and documented.

Responding to that safety recommendation, on December 2018, the Lion Air in coordination with the Lion ground support provider issued Quality Assurance Notice number 102/AQ-AAS/N/XII/2018 to ensure flight crew member configuration were documented properly. In addition, on February 2019, the Lion Air Chief Pilot issued Notice to Pilot number 010/NTP/II/2019 which reminded all pilot to cross check the flight crew member composition with flight attendant and ramp officer.

04.O-2018-35.1
Refer to the CASR Part 91.7 Civil Aircraft Airworthiness and the Operation Manual part A subchapter 1.4.2, the pilot in command shall discontinue the flight when un-airworthy mechanical, electrical, or structural conditions occur.

The flight from Denpasar to Jakarta experienced stick shaker activation during the takeoff rotation and remained active throughout the flight. This condition is considered as un-airworthy condition and the flight shall not be continued.

KNKT recommend ensuring the implementation of the Operation Manual part A subchapter 1.4.2 in order to improve the safety culture and to enable the pilot to make proper decision to continue the flight.

Responding to that safety recommendation, on 21 February 2019, the Lion Air emphasized the OM-part A chapter 1.4.2 through initial training, upgrading (captaincy) program, recurrent and Pilot Proficiency Check (PPC).

Other safety actions taken were as follows:

1. On 14 February 2019, the Operation Training Department issued the CRM training related to crew coordination:
a. Recurrent CRM Lesson Plan
b. Line Training Syllabus

The Operation Training Department also approved the Angkasa Training Center's:

a. ATC/TP/B737/INH INST 01/II/2019 Initial Type Rating Training B737-NG Instructor Lesson Plan

b. Upgrading Syllabus Issue 1 Rev 2 Trainee Syllabus

2. On 25 February 2019, the Operation Training Department issued revision to the training syllabus to enhance flight crew decision making concept during emergency or abnormal situation:

a. B737 Recurrent 5 Ground Class Instructor Lesson Plan number: LA-DO-03-DCOT-134.

b. A330 Recurrent 6 2019 Ground Class Instructor Lesson Plan number: LA-DO-03-DCOT-509.

3. On 25 February 2019, the Operation Training Department issued "B737 Recurrent 5 Ground Class Instructor Lesson Plan" number LA-DO-03-DCOT-134 to improving the flight crew awareness to the aircraft transponder system.

4. On 1 October 2019, implement new training standard and pilot performance review program called "Pilot Performance Enhancement Program" to improve training method and enhance pilot performance to respond KNKT draft recommendation in KNKT draft final investigation report.

5. On 8 October 2019, issued Notice to Pilot number 44/NTP/X/2019 Reminder of MEL Applicability related to enhancing crew decision to comply with requirement for dispatch to respond KNKT draft recommendation in KNKT draft final investigation report.

6. On 9 October 2019, issued Notice to Pilot number 45/NTP/X/2019 DISTRESS and URGENCY MESSAGE Emphasizing flight crew declaration regarding distress message when appropriate to provide awareness to the controller and enabling proper handling to respond KNKT draft recommendation in KNKT draft final investigation report.

7. On 9 October 2019, issued Notice to Pilot number 047/NTP/X/2019 ASR to enhance and detail handing over control procedure to respond KNKT draft recommendation in KNKT draft final investigation report.

8. On 10 October 2019, issued Notice to Pilot number 047/NTP/X/2019 ASR informing synchronization of reporting system definition in the OM-A and SMSM.

9. On 11 June 2019, amended CMM related to the repetitive problem handling policy which aligns with BAT LMPM that amended on 22 July 2019.

Issued revision to the ACL, CAMP and issued Notice to Pilot number 026/NTP/XI/2018 related to PIC duty and responsibility in fault reporting.

Batam Aero Technic

On 08 November 2018, the Batam Aero Technic (BAT) issued Engineering Information to revise Aircraft Flight Manual (AFM) of

Boeing 737-8 (MAX) in accordance with Directorate General of Civil Aviation Airworthiness Directive number 18-11-011-U.

On 11 November 2018, the BAT conducted Angle of Attack installation test to all Boeing 737-8 (MAX) aircraft operated by Lion Air.

Boeing Company

On 6 November 2018, issued Flight Crew Operation Manual Bulletin (OMB) Number TBC-19 with subjected Un-commanded Nose Down Stabilizer Trim Due to Erroneous Angle of Attack (AOA) During Manual Flight Only to emphasize the procedures provided in the runaway stabilizer non-normal checklist (NNC). The detail of the FCOM Bulletin is available on the appendices 5.11.

On 11 November 2018, informed all 737NG/MAX Costumers, Regional Directors, Regional Managers and Boeing Field Service Bases via Multi Operator Messages (MOM) with subject Information – Multi Model Stall Warning and Pitch Augmentation Operation. The detail of the MOM is available on the appendices 5.12.

1. Updates to the MCAS 2. Updates to the 737 MAX Display System 3. Developing changes to the 737 MAX FCOM and certain non-normal procedures.

4. Developing updated detailed computer-based pilot training and other supplemental materials for 737 MAX and MCAS, as requirements above and beyond the instructor-led academic and simulator training that pilots must successfully complete to first qualify to fly the 737 MAX.

Boeing advised the KNKT that a number of actions are being undertaken. Most significantly, Boeing has updated how the MCAS function works and how the AOA Sensor systems on the aircraft interacts with MCAS. Boeing is also in the process of updating crew manuals and training to provide enhanced information that pilots can use to fly the 737 MAX safely.

- The flight control system will now compare inputs from both AOA sensors. If the sensors disagree by 5.5 degrees or more, MCAS will not activate. An indicator on the flight deck display will alert the pilots.

- If MCAS is activated in non-normal conditions, it will only provide one input for each elevated AOA event. There are no known or envisioned failure conditions where MCAS will provide repetitive inputs.

- MCAS can never command more stabilizer input than can be counteracted by the flight crew pulling back on the column. The pilots will continue to always have the ability to override MCAS and manually control the aircraft.
- The Display System will be updated so that all 737 MAX aircraft will have an activated and operable AOA DISAGREE alert and an optional angle of attack indicator.

Collins Aerospace

The safety action taken by Collins Aerospace regards to releasing an update to the CMM revision 10 of CMM 34-12-34 that was released on 2 August 2019, to add robustness testing that included:

. Updates to the test equipment requirements that should allow for easier oversight of equipment equivalency justifications by the MRO.

2. Additions of an Angle Position Indicator verification test that validates modes (resolver vs synchro vs relative) and interconnections.

3. Addition of a resolver output voltage level test that indicates proper transformation ratio of the device (this should identify intermittent resolver outputs at room temperature).

4. Addition of a thermal cycling test for troubleshooting on units that exhibit issues that have altitude (thermal) related on aircraft faults.

AirNav Indonesia Branch Office JATSC

On 4 – 12 February 2019, conducted emergency recurrent training for air traffic controller supervisor on Tower and Approach Control units.
On 27 August 2019, reviewed and amended the JATSC Standard Operation Procedure for Approach Control Services as follows:
- Amended subchapter 6.2.1 to the following:

Old Version
If any report of aircraft instrument malfunction from flight crew might be suspected or classified as an emergency situation

Amended Version
If any report of aircraft malfunction from flight crew might be suspected or classified as an emergency situation

- Added procedure to handle emergency situation as follows:

o. If it is required, controller can request pilot intention during emergency condition with phraseology "REPORT YOUR INTENTION".

Federal Aviation Administration

Post Lion Air accident, actions taken by FAA were:

1. The Federal Aviation Administration (FAA), was assigned as a technical advisor to the US accredited representative, The National Transportation Safety Board (NTSB), to support KNKT Annex 13 accident investigation into Lion Air flight 610. The FAA assigned two individuals who traveled with the NTSB and General Electric (GE). One Accident Investigator and one Flight Test pilot who is type rated in the 737 with 737 Max certification evaluation experience.
2. FAA support is not limited to on-sight activities. The Seattle Aircraft Certification Office has been extensively involved since the accident happened. The FAA participated in the Boeing Safety Review Board process. The Certification office has assigned individuals with expertise in Aircraft systems, flight controls, flight test certification, and aircraft performance to support the Indonesian KNKT. These individuals are continuously working closely with NTSB counterparts.
3. Once an unsafe condition of a possible erroneously high angle of attack (AOA) sensor input is received by the flight control system, there is a potential for repeated nose-down trim commands of the horizontal stabilizer. The FAA immediately began working with Boeing Flight Safety on avenues of communicating the possible unsafe condition to 737-8/-9 operators in the US and Worldwide.
4. The FAA Aircraft Certification Service completed an initial assessment of the Boeing 737-8/-9 aircraft following established transport aircraft continued operational safety risk assessment methodology. The FAA conducted a Corrective Action Review Board (CARB) which produced a fleet risk that required immediate action.
5. On 6 November 2018: The FAA issued a Continued Airworthiness Notification to the International Community (CANIC), notifying foreign civil aviation authorities of our planned airworthiness actions.
6. On 7 November 2018: The FAA issued Emergency Airworthiness Directive (AD) 2018-23-51. The AD was sent to owners and operators of The Boeing Company Model 737-8 and -9 aircraft.
7. On 7 November 2018, The FAA sent Inspectors from the Miramar Flight Standards District Office to Xtra Aerospace to get copies of the repaired AOA P/N 0861FL1 S/N 14488 records, copied those records and provided them to the investigative team.
8. On November 2018: FAA Aircraft Certification Services began working with Boeing to update the MCAS software

a. November and December 2018: FAA simulator evaluations
b. December 2018: FAA review of initial certification planning
9. On January 2019: FAA review of certification plan submittal
a. Changes made through software to the Flight Control Computer Minimize MCAS vulnerability to erroneous AOA signals
b. Limits MCAS command to a maximum of one input for each aircraft high AOA event
c. Limits MCAS maximum command to ensure sufficient handling capability using elevator alone.
d. Changes to the Primary Flight Display System Implementation of the AOA Disagree message
10. On 13 March 2019: FAA grounds 737 MAX once potential relationship between accident flights was established based on Flight profile comparison and Physical evidence from Ethiopia Air accident March 10, 2019.
11. On March 2019: Aircraft Evaluation Group (AEG)/Joint Operations Evaluation Board (JOEB) simulator evaluation of MCAS changes
12. On April and May 2019: FAA initiated international Certification Outreach Sessions
a. 10 outreach sessions conducted;
b. FAA holds conference call with Indonesian DGCA
c. Calls structured to provide common updates to all participating authorities
d. MCAS design changes and certification progress
e. Process for evaluating training changes
f. Outreach efforts will continue as the 737 MAX is returned to service
13. On April 2019: FAA initiates and leads Joint Authorities Technical Review (JATR)
a. Participation of FAA, NASA, and nine Civil Aviation Authorities (CAA)

b. Multi authority review of FAA process and procedures used in the certification of the automated flight control system of the Boeing 737-8/-9 MAX, including human factors and training requirements. The JATR comprises experts from variety of disciplines from the FAA, NASA, and civil aviation authorities of Australia, Brazil, Canada, China, European Union, Indonesia, Japan, Singapore, and the United Arab Emirates.
14. On May 2019: Technical Advisory Board (TAB) initiated consistent of an independent group of FAA and other agency experts not involved in the certification of the Boeing 737 MAX.
a. Participation in the TAB includes: • FAA Experts; Aerospace Engineers (Avionics/Electrical), Flight Standards (FSB & Training), Flight Test Pilot, Chief Scientists (Flight Deck Integration, Software, Flight Simulator Systems)

• Non-FAA Experts; NASA: Flight Controls & System Integration, Volpe: Pilot/Human Factors Chief of Aviation Safety Management Systems, U.S. Air Force: Systems Safety & Integration

b. TAB Activities (In advance of any additional FAA design approvals): • Review the design change and the overall approach to demonstrate compliance with regulatory standards
• Review relevant failure modes that can affect MCAS function and confirm that the design change mitigates hazards
• Review approach for software certification
• Review training program changes
15. May 2019: FAA sponsored the Directors General Aviation Safety Summit
a. Twenty seven countries Including Indonesia, Civil Airworthiness Authorities, EASA, and ICAO were in attendance.
b. FAA presented material on both 737 MAX accidents, a status of the design and certification changes, and the eventual return to service requirements in the United States.
c. FAA is committed to assisting its international partners in the eventual return to service of the 737-8/-9 MAX aircraft.
16. Design Change Validation Efforts
• Transport Canada Certification Authority (TCCA), European Aviation Safety Agency (EASA), the Civil Aviation Authority of China (CAAC), and National Civil Aviation Agency of Brazil (ANAC) are validating authorities evaluating of the 737 MAX design change.
17. FAA is conducting daily phone calls both internally and with Boeing to assist in tracking action items of the 737 MAX design change to eventually return the aircraft to service in the United States. The FAA is committed to assisting CAAs with their own decisions to return the 737 MAX to service.

NTSB Recommendation to FAA

Following the accidents of Lion Air and Ethiopian Airlines, the NTSB participates on both investigations. On 19 September 2019, NTSB issued Safety Recommendation Report titled: Assumptions Used in the Safety Assessment Process and the Effects of Multiple Alerts and Indications on Pilot Performance. The NTSB recommendations to FAA are as follows:
a. Require that Boeing:

(1) Ensure that system safety assessments for the 737 MAX in which it assumed immediate and appropriate pilot corrective actions in response to uncommanded flight control inputs, from systems such as the Maneuvering Characteristics Augmentation System, consider the effect of

all possible flight deck alerts and indications on pilot recognition and response; and

(2) Incorporate design enhancements (including flight deck alerts and indications), pilot procedures, and/or training requirements, where needed, to minimize the potential for and safety impact of pilot actions that are inconsistent with manufacturer assumptions. (A-19-10)

b. Require that for all other US type-certificated transport-category airplanes, manufacturers

(1) Ensure that system safety assessments for which they assumed immediate and appropriate pilot corrective actions in response to uncommanded flight control inputs consider the effect of all possible flight deck alerts and indications on pilot recognition and response; and

(2) Incorporate design enhancements (including flight deck alerts and indications), pilot procedures, and/or training requirements, where needed, to minimize the potential for and safety impact of pilot actions that are inconsistent with manufacturer assumptions. (A-19-11)

c. Notify other international regulators that certify transport-category airplane type designs (for example, the European Union Aviation Safety Agency, Transport Canada, the National Civil Aviation Agency-Brazil, the Civil Aviation Administration of China, and the Russian Federal Air Transport Agency) of Recommendation A-19-11 and encourage them to evaluate its relevance to their processes and address any changes, if applicable. (A-19-12)

d. Develop robust tools and methods, with the input of industry and human factors experts, for use in validating assumptions about pilot recognition and response to safety-significant failure conditions as part of the design certification process. (A-19-13)

e. Once the tools and methods have been developed as recommended in Recommendation A-19-13, revise existing Federal Aviation Administration (FAA) regulations and guidance to incorporate their use and documentation as part of the design certification process, including re-examining the validity of pilot recognition and response assumptions permitted in existing FAA guidance. (A-19-14)

f. Develop design standards, with the input of industry and human factors experts, for aircraft system diagnostic tools that improve the prioritization and clarity of failure indications (direct and indirect) presented to pilots to improve the timeliness and effectiveness of their response. (A-19-15)

Once the design standards have been developed as recommended in Recommendation A-19-15, require implementation of system diagnostic tools on transport-category aircraft to improve the timeliness and effectiveness of pilots' response when multiple flight deck alerts and indications are present. (A-19-16)

Directorate General of Civil Aviation

1. Instructed to all Boeing 737-8 (MAX) to conduct special airworthiness inspection to all Boeing 737-8 (MAX);
2. DGCA performed Special Audit to Lion Air and Batam Aero Teknik from 30 October 2018 to 2 November 2018;
3. DGCA performed Ramp Check to all Boeing 737-8 (MAX) operated by Lion Air which later on continued with other types of aircraft.
4. On 8 November 2018, DGCA issued Airworthiness Directive (AD) number 18-11-011-U which referred to the FAA Emergency Airworthiness Directive number 2018-23-51 and Boeing FCOM Bulletin TBC-19;
5. On 15 November 2018, DGCA issued Safety Circular number SE. 39 of 2018 related to the implementation of Airworthiness Directive number 18-11-011-U;
6. On 21 and 22 November 2018, DGCA performed implementation evaluation of the AD no. 18-11-011-U in simulator Boeing 737-8 (MAX) in Singapore;
7. On 13 December 2018, DGCA issued Safety Circular number SE. 45 of 2018 related to the regulation of occurrence reporting to aircraft operators;
8. On 13 December 2018, DGCA issued Safety Circular number SE. 46 of 2018 related to handling of repetitive problem;
9. On 12 March 2019, DGCA issued *Temporary Grounded* to Boeing 737-8 (MAX) operated by Indonesia aircraft operator;
10. On 22 October 2019, DGCA issued Safety Circular number SE.013 of 2019 related to procedure of hand over control to respond KNKT draft recommendation in KNKT draft final investigation report.

CHAPTER 10

SAFETY RECOMMENDATIONS

The KNKT acknowledges the safety actions taken by Lion Air, Batam Aero Technic, FAA, and DGCA. KNKT considered that the safety actions were relevant to improve safety, and encourage the implementation of these safety actions.

KNKT identified safety issue remains to be considered therefore, the KNKT issued safety recommendations to address safety issues identified in this report.

Lion Air

04.O-2018-35.3
The investigation found Lion Air manuals did not updated in timely manner and the content have several inconsistencies, incompleteness, and unsynchronized procedures.
Therefore, KNKT recommends establishing system to ensure the company manuals are updated in timely manner.

04.O-2018-35.4
The investigation considered that the duration of hazard identification topic on the SMS training syllabus was insufficient. This may reduce the ability of employees to define and report a hazard. Consistently, the Lion Air safety report on December 2018 mostly consisted of occurrence report and only about five percent of hazard report.

Therefore, KNKT recommends that Lion Air review the SMS training material and the duration of training.

04.O-2018-35.5
The LNI043 flight that experienced multiple malfunctions were considered caused or could have caused difficulties in controlling the aircraft. According to the ICAO Annex 13, CASR part 830 and OM-part A, the flight is classified as serious incident which required investigation by the KNKT in accordance with the Aviation Law Number 1 of 2009 and Government Decree Number 62 of 2013.

Therefore, KNKT recommends that Lion Air improve their hazard report management enabling identifying the hazard and provides proper mitigation.

Batam Aero Technic

04.O-2018-35.6
The AOA installation test was performed by the engineer in Denpasar using the alternative method as described in the Boeing AMM. The AOA values indicated on the SMYD computer during the test were not recorded as required by the BAT LMPM. Without the recorded value, the success of the installation test could not be determined.

Therefore, KNKT recommends emphasizing engineers to record test values as required by the BAT LMPM.

04.O-2018-35.7
The investigation found BAT manuals did not updated in timely manner and the content have several inconsistencies, incompleteness, and unsynchronized procedures.

Therefore, KNKT recommends establishing system to ensure the company manuals are updated in timely manner.

04.O-2018-35.8
The OMF has the history page which contains record of the aircraft problems which can be utilized as a source for aircraft problem monitoring. The BAT has not utilized the OMF information as the source of aircraft problem monitoring.

Therefore, KNKT recommends that Batam Aero Technic establish policy and procedure of handling OMF.

AirNav Indonesia

04.A-2018-35.9
The flight crew of LNI610 asked to the TE controller of the aircraft altitude detected on the ATC radar display which might be an effort to

obtain other source of information. The asking of aircraft altitude to the controller will not get any additional information as the ATC radar display is received data from aircraft transponder which transmitting the cockpit indications.

Therefore, KNKT recommends providing information to the flight crew that the altitude indication on the ATC radar display was repeating data from the aircraft.

Xtra Aerospace

04.O-2018-35.10

After Xtra Aerospace repair of the accident AOA sensor in November of 2017, the sensor was installed on the PK-LQP aircraft on left side position during the maintenance activity in Denpasar on 28 October 2018. On the subsequent flight, a 21-degree difference between left and right AOA sensors was recorded on the DFDR, commencing shortly after the takeoff roll was initiated. This immediate 21- degree delta indicated that the AOA sensor was most likely improperly calibrated at Xtra.

As noted, utilization of the Peak Model SRI-201B API by Xtra Aerospace for the test and calibration of the 0861FL1 AOA sensor should have required a written procedure to specify the proper position of the REL/ABS switch.

Therefore, KNKT recommends emphasizing the implementation of a company manual including equivalency assessment, training and written procedure, to ensure component being repaired are properly maintained.

Boeing Company

04.M-2018-35.11

During the accident, multiple alerts and indications occurred which increased flight crew's workload. This obscured the problem and the flight crew could not arrive at a solution during the initial or subsequent automatic AND stabilizer trim input, such as performing the runaway stabilizer procedure or continuing to use electric trim to reduce column forces and maintain level flight.

Therefore, KNKT recommends that the aircraft manufacture to consider the effect of all possible flight deck alerts and indications on flight crew recognition and response; and incorporate design, flight crew procedures, and/or training requirements where needed to minimize the potential for flight crew actions that are inconsistent with manufacturer assumptions.

04.M-2018-35.12

During certification phase, compliance was demonstrated by flight test pilots which normally have exceptional skill and experience. Flight test pilots generally have more knowledge about the aircraft design characteristics than normal pilots. This level of competence usually cannot be translated to most pilots. However, test pilots are trained to replicate the average flight crew. The Aircraft Evaluation Group pilots, who have an operational flying background, also evaluate the aircraft during the certification phase. These pilots establish the pilot type rating, training, checking and currency requirements as part of the Flight Standardization Board (FSB) process. The FSB process also utilizes airline line pilots to help ensure the requirements are operationally representative. The FAA and OEMs should re-evaluate their assumptions for what constitutes an average flight crew's basic skill and what level of systems knowledge a 'properly trained average flight crew' has when encountering failures.

Therefore, KNKT recommends that Boeing include a larger tolerance in the design is required to allow operability by a larger population of flight-rated pilots.

04.M-2018-35.13

During the accident flight, the DFDR recorded a control force of 103 lbs., after repetitive MCAS activation was responded with the FO had responded with inadequate trim to counter MCAS. At this point, the flight crew was unable to maintain altitude.

Therefore, KNKT recommends that Boeing and the FAA more closely scrutinize the development and certification process for systems whose malfunction has the ability to lead to loss of control of the airplane.

04.M-2018-35.14

The flight crew should have been provided with information and alerts to help them understand the system and know how to resolve potential issues. Flight crew procedures and training should be appropriate.

Therefore, KNKT recommends to Boeing to develop the guidance for the criteria of information which should be included in flight crew and engineer's manuals.

04.M-2018-35.15

The aircraft should have included the intended AOA DISAGREE alert message functionally, which was installed on 737 NG aircraft. Boeing and the FAA should ensure that new and changed aircraft design are properly described, analyzed, and certified.

Therefore, KNKT recommends to Boeing that they ensure that certified and delivered airplanes have intended system functionality.

04.M-2018-35.16

The IFIM tasks of "ALT DISAGREE" and "IAS DISAGREE" are duplicated on the leak test in step (3) and (4) as they are referring to the

same AMM tasks. This repetition was inefficient and did not contribute to problem solving.
Therefore, KNKT recommends the IFIM tasks sequence are reviewed to ensure they are effective.

Directorate General of Civil Aviation (DGCA)

04.R-2018-35.17
The investigation revealed several deviations from standard procedures such as departure with unserviceable airworthiness items and inappropriate implementation of maintenance procedure.
Therefore, KNKT recommends improving oversight to ensure implementation of the standard procedure.

04.R-2018-35.18
Several manuals were found inconsistence, inapplicable, the content did not include duty and responsibility of significant personnel and not updated in timely manner.
Therefore, KNKT recommends improving the oversight of the manual to ensure the manuals are conform to the standard and updated in timely manner.

04.R-2018-35.19
The investigation considered that the duration of hazard identification topic on the SMS training syllabus was insufficient. This may reduce the ability of employees to define and report a hazard. Consistently, the Lion Air safety report on December 2018 mostly consisted of occurrence report and only about five percent of hazard report.
Therefore, KNKT recommends that the Directorate General of Civil Aviation review operators SMS training material and the duration of training to ensure adequacy of SMS implementation.

Federal Aviation Administration (FAA)

04.R-2018-35.20
In the accident flight, the system malfunction led to erroneous information that initiated a series of events that were not correctly recognized and responded to by the flight crew.
This exposed issues that were not identified if FAR 25.1302 and 25.1309 were each considered separately in which system malfunction was followed by flight crew limitation in identifying and mitigating the problem. There could be a potential gap between the two requirements when system malfunction is followed by crew fallibility.

Therefore, KNKT recommends to review the requirements of the applicable FARs to consider any issue that may be overlooked when the requirements are considered separately.

04.R-2018-35.21
In the accident flight, the system malfunction led to a series of aircraft and flight crew interactions which the flight crew did not understand or know how to resolve. It is the flight crew response assumptions in the initial design process which, coupled with the repetitive MCAS activations, turned out to be incorrect and inconsistent with the FHA classification of Major.

Therefore, the KNKT recommends that the FAA review their processes for determining their level of involvement (degree of delegation) and how changes in the design are communicated to the FAA to ensure an appropriate level of review.

04.R-2018-35.22
The absence of equivalency assessment required by Xtra Aerospace procedure and unavailability of procedure was not detected by the FAA. This indicated inadequacy of the FAA oversight.

Therefore, KNKT recommends that the FAA improves the oversight to Approved Maintenance Organization (AMO) to ensure the processes within the AMO are conducted in accordance with the requirements.

04.R-2018-35.23
During the accident flight, the DFDR recorded a control force of 103 lbs., after repetitive MCAS activation was responded with the FO had responded with inadequate trim to counter MCAS. At this point, the flight crew was unable to maintain altitude.

Therefore, KNKT recommends that Boeing and the FAA more closely scrutinize the development and certification process for systems whose malfunction has the ability to lead to loss of control of the airplane.

04.R-2018-35.24
During the accident and previous LNI043 flights, the flight crew initially responded in the same way, by pulling back on the control column. However, they did not consistently trim out the resulting column forces as had been assumed. As a result Boeing assumption was different from the flight crew behavior and reaction time in responding to MCAS activation.

Therefore, the KNKT recommends that the FAA work with international regulatory authorities to review assumptions on flight crew behavior used during design and revise certification processes to ensure assumptions used during the design process are validated.

04.R-2018-35.25

The flight crew should have been provided with information and alerts to help them understand the system and know how to resolve potential issues. Flight crew procedures and training should be appropriate.

Therefore, KNKT recommends to the FAA work with international regulatory authorities to review the guidance for the criteria of information which should be included in flight crew and engineer's manuals.

04.R-2018-35.26

The aircraft should have included the intended AOA DISAGREE alert message functionally, which was installed on 737 NG aircraft. Boeing and the FAA should ensure that new and changed aircraft design are properly described, analyzed, and certified.

Therefore, KNKT recommends to Boeing and the FAA that they ensure that certified and delivered airplanes have intended system functionality.

04.R-2018-35.27

The aircraft was equipped with an airframe-mounted low frequency underwater locator beacon (ULB) which operated at a frequency of 8.8 kHz. The beacon is included in ICAO standards. The purpose of the beacon is to aid in the location of submerged aircraft. During the search phase, multiple surveys were conducted to detect a signal at 8.8 kHz, however no such signals were detected in the area where wreckage was recovered.

The beacon was mounted on the forward side of the nose pressure bulkhead. Most of the preferred installation locations could not be used because they proved to be incompatible with EASA and FAA Non-Rechargeable Lithium Battery certification requirements or they did not meet the ICAO empennage and wings exclusion.

Therefore, KNKT recommends to the FAA work with international regulatory authorities to review the requirements for installation of Non-Rechargeable Lithium Battery certification requirements.

APPENDICES

APPENDIX 1: NTSB SYSTEM SAFETY AND CERTIFICATION SPECIALIST'S REPORT

NATIONAL TRANSPORTATION SAFETY BOARD
OFFICE OF AVIATION SAFETY
WASHINGTON, D.C. 20594
August 21, 2019

SYSTEM SAFETY AND CERTIFICATION SPECIALIST'S REPORT

NTSB ID No.: DCA19RA017

A. ACCIDENT:

Operator:	Lion Mentari Airlines (Lion Air)
Location:	Jakarta, Indonesia
Date:	October 28, 2018
Aircraft:	Boeing 737-8 (MAX), Registration PK-LQP

B. SUMMARY:

On October 29, 2018, PT Lion Mentari Airlines (Lion Air) flight 610, a Boeing 737 MAX 8, PK-LQP, crashed in the Java Sea shortly after takeoff from Soekarno-Hatta International Airport, Jakarta, Indonesia. The flight was a scheduled domestic flight from Jakarta to Depati Amir Airport, Pangkal Pinang City, Bangka Belitung Islands Province, Indonesia. All 189 passengers and crew on board died, and the aircraft was destroyed. The National Transportation Safety Committee of Indonesia is leading the investigation (The preliminary report on this accident can be found at https://reports.aviation-safety.net/2018/20181029-0_B38M_PK-LQP_PRELIMINARY.pdf)

C. 737 MAX AND THE NEED FOR MCAS:

The Boeing 737-8 (MAX) is a derivative of the 737-800 model and is part of the 737 MAX family (737 MAX 7, 8, and 9 [27]). The 737 MAX incorporated the CFM LEAP-1B engine, which has a larger fan diameter and redesigned engine nacelle compared to engines installed on the 737 Next Generation (NG) family. Because the 737-8 is a derivative of the 737-800 model, its certification basis, which was established per 14 CFR 21.101 Changed Product Rule, required Boeing to demonstrate compliance with Amendment 25-136 for significant areas of change at the product level and those areas affected by the significant product level change.

During the preliminary design stage of the Boeing 737-8 (MAX), Boeing tests and analysis revealed that the addition of the LEAP-1B engine and associated nacelle changes produced an aircraft nose-up pitching moment when the aircraft was operating at high angles of attack (AOA) and mid Mach numbers. This nose-up pitching moment was deemed likely to affect the stick force per g (FS/g) characteristics required by FAR 25.255 and the controllability and maneuverability requirements of FAR 25.143(f). After the study of various options for addressing this issue, Boeing implemented aerodynamic changes as well as a stability augmentation function called the Maneuvering Characteristics Augmentation System (MCAS), as an extension of the existing Speed Trim System (STS), to improve aircraft handling characteristics and decrease pitch-up tendency at elevated angles of attack.

As the development of the Boeing 737-8 (MAX) progressed, the MCAS function was expanded to low Mach numbers. MCAS is designed to function only during manual flight (autopilot not engaged), with the aircraft flaps up, at an elevated AOA.

D. SPEED TRIM & MCAS DESCRIPTION:

To ensure that the 737-600/700/800/900 (737 NG) family of aircraft met the certification requirements for longitudinal static stability (speed stability), the aircraft incorporated a Speed Trim System (STS) to augment the basic aircraft speed stability during certain low speed, high thrust flight conditions by moving the horizontal stabilizer during manual flight (autopilot is not engaged). For the 737 NG family of aircrafts, the Speed Trim System included the Speed Trim Function. The STS was carried over to the 737-7/-8/-9 (737 MAX) family of aircraft. Additionally, on 737 MAX aircraft, the MCAS function was added to the STS to address the pitch characteristics described above.

D.1 Speed Trim Function:

The Speed Trim function, which is implemented as a control law within the flight control computer (FCC[28]), commands incremental stabilizer trim through the automatic trim control system circuitry. There are two different stabilizer trim rates depending on whether position of the flaps[29].

A schedule determines the desired incremental stab deviation from the last trimmed position as a function of airspeed and flap position.

According to the Enhanced Digital Flight Control System (EDFCS) system safety analysis (SSA), the worst-case failure mode of the Speed Trim function was considered to be a runaway of the horizontal stabilizer trim actuator (HSTA) as a result of sensor or FCC failures, or FCC-to-stabilizer trim motor (STM) wiring failures. The SSA indicated that during the runaway, the flight crew is able to detect the fault by noticing the continuous running of the trim mechanical wheels in the flight deck, or by the change in column force necessary to maintain pitch attitude, or through change in aircraft pitch attitude. The SSA indicated that the flight crew compensates for the runaway through:

- column input in the direction opposing the uncommanded trim until activation of the column activated trim cutout switches, or
- activation of the main electric trim by either flight crew in a direction opposing the uncommanded motion, which overrides the FCC commanded trim runaway, or
- moving the guarded stabilizer trim cutout switches[30] located on the aisle stand to the CUTOUT position, or restraining the stabilizer trim wheel,
- Speed/ Stab Trim runaways are limited by the inherent stabilizer trim motor rate and column actuated trim cut-out switches. Sufficient means are available for the flight crew to maintain control and recover from the runaway[31].

D.2 MCAS Functional – Detailed Description:

The MCAS is a function within the Speed Trim System and, when activated, moves the stabilizer during non-normal flaps up, high angle of attack maneuvers to provide a desirable increase in stick force gradient and a reduced pitch up tendency. Similar to the Speed Trim Function, the MCAS function is also a flight control law[32] contained within each of the two FCCs. MCAS is only active in the master FCC for that flight.

As originally delivered, the MCAS became active during manual, flaps-up flight (autopilot not engaged) when the AOA value received by the master FCC exceeded a threshold based on Mach number. When activated, the MCAS provided a high rate automatic trim command to move the stabilizer AND. The magnitude of the AND command was based on the AOA and the Mach. After the non-normal maneuver that resulted in the high AOA, and once the AOA fell below a reset threshold, MCAS would move the stabilizer ANU to the original position and reset the system. At any time, the stabilizer inputs could be stopped or reversed by the pilots using their yoke-mounted electric stabilizer trim switches, which also reset the system after a 5 second delay.

The latter behavior is based on the assumption that flight crews use the trim switches to completely return the aircraft to neutral trim. In the FCC software version current at the time of the accident, if the original elevated AOA condition persists for more than 5 seconds following an MCAS flight control law reset, the MCAS flight control law will command another stabilizer nose down trim input (with the magnitude based on the AOA and Mach sensed at that time).

On all 737 models, column cutout switches interrupt stabilizer commands, either from the autoflight system (e.g. FCC) or the electric trim switches in a direction opposite to elevator command. On the 737NG and MAX, two column cutout switching modules, one for each control column, are actuated when the control columns are pushed or pulled away from zero (hands off) column position. When actuated, the column cutout switching modules interrupt the electrical signals to the stabilizer trim motor that are in opposition to the elevator command.

The MCAS function requires the stabilizer to move nose down in opposition to the column commands when approaching high angles of attack. To accommodate MCAS, the column cutout function in the first officer's switching module was modified to inhibit the aft column cutout switch while MCAS is active, allowing aircraft nose-down (AND) stabilizer motion with aircraft nose-up (ANU) column input. Once MCAS is no longer active, the normal column cutout function in the stabilizer nose down direction is re-instated

E. FUNCTIONAL HAZARD ASSESSMENT AND REQUIREMENTS GENERATION:

E.1 Functional Hazard Assessment:

A functional hazard assessment (FHA) is a systematic examination of a system's functions and purpose, and it typically provides the initial, top-level assessment of a design and addresses the operational vulnerabilities of the system function. The FHA is therefore typically used to establish the safety requirements that guide system architecture design decisions. An FHA evaluates what would occur (the "hazard" in FHA) if the function under question was lost or malfunctioned and classifies the severity of that effect. An FHA is conducted early in the design and development cycle to identify hazards and classify them by severity, beginning at the aircraft level and working down to individual systems.

Federal Aviation Administration (FAA) Advisory Circular AC 25.1309-1A, dated June 21, 1988 and SAE ARP4761 define the severity classes that are used to classify the effect of loss or malfunction as part of an FHA. AC 25.1309-1A defines the following three severity classes: catastrophic, major and minor, with corresponding acceptable probabilities of extremely improbable (1E-9) or less per flight hour), improbable (1E-5 or less), and no worse than probable (1E-3). European regulations (originally JAR and now EASA) include an additional category: hazardous, which falls between catastrophic and major and has an associated acceptable probability of 1E-7 or less. The differences among the classes are associated with effects on the aircraft, occupants, and crew.

To begin an FHA, engineering judgment is used to identify the failure conditions which require evaluation. According to the FHA sections[33] of Boeing's 737 NG/MAX Stabilizer Trim Control System Safety Analysis, (Reference section H.2.2 of this report), performance analyses and piloted simulations were accomplished as needed to help define the hazard categories for the identified conditions. shows the criticality categories used in developing the FHA and the corresponding minimum acceptable probabilities of occurrence. The failure conditions defined by the FHA provide the basis for the top-level events analyzed by the Fault Tree Analysis (FTA) to demonstrate compliance with FAR 25.671(c)(2) and 25.1309(b)(1). A fault tree analysis was performed on each failure condition determined to be either Catastrophic or Hazardous. Additionally, Major events are included in the FHA for reference, per FAA/JAA request.

As part of the MCAS development phase, in late 2012, Boeing performed a preliminary functional hazard assessment[34] of MCAS using piloted simulations in their full motion Engineering Flight Simulator. Several hazards were assessed at that time, however, this section of the report will focus only on the following two hazards: uncommanded MCAS operation up to its maximum authority (0.6 degrees of aircraft nose down stabilizer) and uncommanded MCAS operation equivalent to a three (3) second stabilizer trim runaway [35]. To perform these simulator tests, Boeing induced a stabilizer trim input that would simulate the stabilizer moving at a rate and duration consistent with the MCAS function. Using this method to induce the hazard resulted in the following: motion of the stabilizer trim wheel, increased column forces, and indication that the aircraft was moving nose down. Boeing indicated to the NTSB that this evaluation was focused on the pilot response to uncommanded MCAS operation, regardless of underlying cause. Thus, the specific failure modes that could lead to uncommanded MCAS activation, such as an erroneous high AOA input to the MCAS, were not simulated as part of these functional hazard assessment validation tests. As a result, additional flight deck effects (such as IAS DISAGREE and ALT DISAGREE alerts and stick shaker activation) resulting from the same underlying failure (for example, erroneous AOA) were not simulated and were not documented in the stabilizer trim and autoflight safety assessment reports reviewed by the NTSB.

The FHA evaluations were conducted by Boeing in their Engineering Cab using FAA guidance regarding pilot response to flight control failures requiring trim input that is contained in FAA Advisory Circular AC25.7C[36]. In particular, Boeing uses the following assumptions in its flight controls FHAs:

- Uncommanded system inputs are readily recognizable and can be counteracted by overriding the failure by movement of the flight controls in the normal sense by the flight crew and do not require specific procedures.

- Action to counter the failure shall not require exceptional piloting skill or strength.

- The pilot will take immediate action to reduce or eliminate increased control forces by re-trimming or changing configuration or flight conditions.

- Trained flight crew memory procedures shall be followed to address and eliminate or mitigate the failure.

Boeing advised that these assumptions are used across all Boeing models when performing functional hazard assessments of flight control systems and that these assumptions are consistent with the requirements contained in 14 CFR 25.671 & 25.672 and within the guidance contained in FAA Advisory Circular (AC) 25-7C for compliance evaluation of 14 CFR 25.1433[9].

In March 2016, Boeing determined that MCAS should be revised to improve wings-level, flaps up, low Mach stall characteristics and identification. The MCAS was revised such that depending on AOA, it would be capable of commanding incremental stabilizer to a maximum of 2.5 degrees at low Mach decreasing to a maximum of 0.65 degrees at high Mach.

The requirements document also indicated that the preliminary functional hazard assessments of MCAS were re-evaluated by pilot assessments in the motion simulator and by engineering analysis and determined to have not changed in hazard classification as a result of the increase in MCAS authority to 2.5 degrees.

When assessing unintended MCAS activation in the simulator for the FHAs, the function was allowed to perform to its authority and beyond before pilot action was taken to recover. Failures were able to be countered by using elevator alone. Stabilizer trim was available to offload column forces, and stabilizer cutouts were available but not required to counter failures. This was true both for the preliminary FHAs performed in 2012 and for the reassessment of the FHAs in 2016.

In a 2019 presentation to the NTSB, Boeing indicated that the MCAS hazard classification of "Major" for uncommanded MCAS function (including up to the new authority limits) in the Normal flight envelope were based on the following conclusions:

- Unintended stabilizer trim inputs are readily recognized by movement of the stabilizer trim wheel, flight path change or increased column forces.
- Aircraft can be returned to steady level flight using available column (elevator) alone or stabilizer trim.
- Continuous unintended nose down stabilizer trim inputs would be recognized as a Stab Trim or Stab Runaway failure and procedure for Stab Runaway would be followed.

Boeing also indicated that as part of the development process, although not formally part of the FHA analysis, engineering personnel and test pilots discussed the scenario of repeated uncommanded MCAS activation due to erroneously high AOA and considered whether a system redesign was necessary to address this issue. As part of this discussion, they discussed the combined flight deck effects (including stick shaker activation, among others), but determined that no redesign was necessary. This conclusion was based in part on the assumption that each activation would be recognized and immediately trimmed out, which is consistent with the regulatory guidance in AC 25-7C that a pilot will take immediate action to trim out reduce or eliminate high control forces by re-trimming or changing configuration or flight conditions.

E.1.1 Requirements Generation and Traceability:

Based on the MCAS pilot assessments using the Engineering Flight Simulator, several system and safety requirements were generated. An NTSB review of these requirements found one requirement related to the probability of an MCAS system hardover. The requirement stated: "The probability of a system hardover, oscillatory failure, and loss of function shall be commensurate with the hazard levels identified by the FHA, which were determined by Pilot simulator assessments of the MCAS failure modes. As previously stated, unintended MCAS operational FHA events were assessed as "Major" in the normal flight envelope, with a corresponding required probability of 1E-5.

The MCAS function is a control law (software) contained within the Flight Control Computer (FCC), which was developed by Rockwell Collins Inc to meet the design specifications contained within a Specification Control Drawing (SCD) provided to them by Boeing[37]. The SCD covers the design, fabrication, performance, qualification, and functional testing requirements for the Enhanced Digital Flight Control System (EDFCS) for use on the Boeing 737. An NTSB review of the SCD revealed that requirements for MCAS were first added to the document at Revision G, dated July 28, 2014.

On December 23, 2015, Boeing released an internal document titled "Engineering Authorization for Incorporation of EDFCS Problem Reports B-1740" to transmit requested changes, safety requirements, into the EDFCS SCD. Of the six new safety requirements, two of them were related to MCAS; One of the safety requirements (3.1.1.5.3.1.1-A) included an upper limit to "The probability of the FCC producing an erroneous flaps up/down discrete output or an erroneous MCAS Engage discrete output without detection." This requirement was derived from the above-mentioned FHA result that unintended MCAS operation have a probability of less than 1E-5.

An NTSB review of the EDFCS SCD revealed that the MCAS safety requirement 3.1.1.5.3.1.1-A was added to the SCD per Boeing document "B-1740" at Revision J, dated November 3, 2016.

An NTSB review of a December 09, 2016 Rockwell Collins document titled "EDFCS FCC-730 P10.0 Requirement Verification Matrix" was conducted. This document included a "traceability matrix" table that identified the incremental requirements that were changed/added/deleted for the EDFCS FCC-730 P10.0 software development. The document indicated that the traceability matrix had been reviewed by Rockwell and their review found that the requirements affected by the EDFCS FCC-730 P10.0 software development have been correctly allocated, implemented, and verified. The NTSB review of the "traceability matrix" table found that it included all of the safety requirements that were added to the SCD per Boeing document "B-1740, including the MCAS safety requirement 3.1.1.5.3.1.1-A. According to Boeing, the safety requirement would be covered in the EDFCS system safety assessment. A review on the Boeing EDFCS system safety assessment found that the MCAS safety requirement 3.1.1.5.3.1.1-A was addressed.

F CERTIFICATION:

In Title 14, Code of Federal Regulations (CFR) United States of America Part 21, the Federal Aviation Administration (FAA) is responsible for certifying aircraft. The certification basis is usually established based on the aircraft configuration and functionalities and any special conditions that deemed necessary. For Boeing 737-8 MAX the certification basis is mainly based on the FAR Part 25. Boeing is responsible to show compliance with the requirements set in the certification basis using a proper and standard procedure[38].

F.1 Type Certification Process and Overview:

The FAA is responsible for prescribing minimum standards required in the interest of safety for the design, material, construction, quality of work, and performance of aircraft, aircraft engines, and propellers (Ref. 49USC44701). Product certification[39] is a regulatory process administered by the FAA to ensure that an aircraft manufacturer's product complies with Federal Aviation Regulations (FAR). Successful completion of the certification process enables the FAA to issue a type certificate (TC) or an amended type certificate (ATC). To obtain a TC or an ATC, the manufacturer must demonstrate to the FAA that the aircraft or product being submitted for approval complies with all applicable regulations. The FAA determines whether or not the applicant has met its responsibility to show compliance to the applicable regulations.

The Federal regulations that apply to type certification of transport-category aircraft are 14 CFR Part 21, 25, 26, 33, 34, and 36. The Part 25 regulations are those concerned with the airworthiness standards for transport-category airplanes and are organized into subparts A through G. Because regulations are continuously evolving, each aircraft is assigned a type certification basis that is established by the FAA based on the regulations in effect on the date of application. These regulations represent the minimum standards for airworthiness; an applicant's design may exceed these standards and the applicant's tests and analyses may be more extensive than required by regulation. The specific applicable regulatory requirements and how compliance will be demonstrated is documented in an FAA accepted certification plan.

F.2 Certification Guidance

FAA Order 8110.4C, titled "Type Certification", prescribes the responsibilities and procedures the FAA must follow to certify new civil aircraft, aircraft engines, and propellers, or changes thereto, as required by 14 of the Code of Federal Regulations (CFR) Part 21. This order is primarily written for internal use by the FAA, its designees, and delegated organizations. The order provides procedures and policy for the type certification of products and, unless stated otherwise, the type certification process in this order applies to all U.S. TCs, including amended TCs.

F.3 Typical Certification Process

FAA Order 8110.4C contains a section that presents a high-level flow diagram of the certification events that typically make up the life cycle an aircraft. The diagram is meant to explain the type certification process, not to dictate precisely how the project should flow. Although the model shows the proper sequence of events for certificating a product, the various aspects of the project generally progress through the process at different times and at different rates. The model divides the product's type certification life cycle into phases based on *The FAA and Industry Guide to Product Certification*. For each of the certification events identified on the flow diagram, the Order also provides information describing each event, identifies expectations and develops specific interface procedures between the applicant and the FAA.

During a meeting with the NTSB[40], the FAA provided a high-level overview of the certification process for an amended type design program. The briefing indicated that the applicant would start by conducting familiarization briefings and submitting the following to the FAA: a certification project notification (CPN), a program notification letter (PNL) and a master certification plan (MCP). These documents detail the changes and identify the regulatory requirements and policies that are applicable; they also identify areas of change associated with the FAA airworthiness directives. As part of the overview, the FAA provided a high-level flow diagram of the certification events that contained similar information as the diagram within Order 8110.4c.

During a meeting with the NTSB[41], the FAA provided the investigation team with a list (Reference table 3) showing a timeline for when certain Boeing 737-8 (MAX) certification events occurred.

F.4 FAA Certification Office

The FAA has 10 aircraft certification offices (ACO) which are responsible for approving the design certification of aircraft, aircraft engines, propellers, and replacement parts for those products. There are also specialized certification offices which include the Engine Certification Office (ECO), the Military Certification Office (MCO), the Boeing Aviation Safety Oversight Office (BASOO), and the Delegation Systems Certification Office (DSCO). The BASOO is the FAA's certification office specifically assigned to provide oversight of the certification of Boeing products. It is located in Seattle Washington. BASOOs' responsibilities include oversight of Boeing's Organization Designation Authorization (ODA), involvement in certification of safety critical areas as well as novel and unusual designs and assisting foreign Civil Aviation Authorities (CAAs) in validation of Boeing products. The BASOO was responsible for the certification oversight and approval for the 737 MAX.

F.5 Certification Basis

According to Type Certificate Data Sheet[42] (TCDS) A16WE, revision 64, dated October 10, 2018, Boeing applied for a transport category amended type certificate (ATC) for the 737-8 aircraft on June 30, 2012. The ATC was approved on March 8, 2017. The Boeing 737-8 aircraft was added as the most recent model in a series of derivative models (or "changed aeronautical products") that were approved and added to the Boeing type certificate (TC), originally issued for the Boeing 737-100 on December 15, 1967.

45 A Type Certificate Data Sheet (TCDS) is a formal description of the aircraft, engine or propeller. It lists limitations and information required
The applicable certification basis for the 737-8 aircraft is Title 14, Code of Federal Regulations (14 CFR) part 25 as amended by Amendments 25-0 through 25-137, plus amendment 25-141 with exceptions permitted by 14 CFR 21.101.

F.6 Certification Basis for Changed Aviation Products

The certification basis for changed aeronautical products allows an aircraft manufacturer to introduce a derivative model as a design update on a previously certificated aircraft and add the changed product onto an existing TC. The FAA approves such changes if it finds that the changes are not significant enough to warrant application for a new TC. This process enables a manufacturer to introduce derivative aircraft models without having to resubmit the entire aircraft design for certification review. The manufacturer can use the results of some of the analyses and testing from the original type certification to demonstrate compliance, in which case the regulations that were in effect on the date of the original TC apply.

Title 14 CFR 21.101, Subpart D, specifies the requirements for demonstrating airworthiness compliance for changed aeronautical products. The current revision of 14 CFR 21.101, amendment 21.92, which became effective on April 16, 2011, states that an application for a changed aeronautical product to be added to a TC "must show that the changed product complies with the airworthiness requirements applicable to the category of the product in effect on the date of the application." This regulation is more specific than previous revisions regarding what can be used from the original certification basis in an application for a derivative model involving a major change.

On April 25, 2003, the FAA issued FAA Order 8110.48, *How to Establish the Certification Basis for Changed Aeronautical Products*, which provides the procedures that the FAA and its designees utilize for determining the certification basis for changes to type certificated products including changes made through an amended Type Certificate which was the method utilized for the 737 MAX. The handbook refers to FAA Advisory Circular 21.101-1, *Establishing the Certification Basis of Changed Aeronautical Products*, which contains an acceptable means, but not the only means, to comply with 14 CFR 21.101. On July 21, 2017, this Order 8110.48 was cancelled and replaced by Order 8110.48A.

G SYSTEM SAFETY ASSESSMENT PROCESS - GENERAL:
G.1 Overview

The process for developing and certifying a safety-critical system must provide assurance that all significant single failure conditions have been identified and that all combinations of failures which lead to hazardous or catastrophic aircraft level effects have been considered and appropriately mitigated. Aircraft manufacturers provide this assurance through their safety assessment processes.

The safety assessment process is divided into two parts; the aircraft level safety assessment and the individual system safety assessments. The aircraft safety assessment assures the robustness of the overall aircraft system design that implements the required aircraft functions. The individual system safety assessments assure the system designs meet their safety requirements and support the aircraft level safety assessment.

The aircraft assessment process begins by identifying the aircraft functions and determining which aircraft functions are required for continued safe flight and landing. A Functional Hazard Assessment (FHA) is performed on the functions required for safe flight and landing to identify potentially catastrophic and hazardous failure conditions. For each failure condition, the aircraft architecture (i.e. systems) which implements the function is identified and the high-level system failure conditions are determined. An engineering assessment is performed to verify system failure conditions are being addressed by the individual systems.

Safety assessments are conducted by the applicant, and its suppliers, and are reviewed and approved by the FAA. The safety assessment process is outlined in AC 25.1309-1A and described in detail in SAE ARP4761. Although the safety assessment process outlined in the AC is not mandatory, the AC documents an established means, but not the only means, for an applicant to show compliance to the regulations. An applicant who chooses not to conduct safety assessments must demonstrate compliance in another way, which would have to be FAA-approved.

H CERTIFICATION OF THE MCAS IMPLEMENTATION AND FUNCTION:

H.1 Certification Plans:

H.1.1 Certification Plan Guidance:

When Boeing submitted its application for the 737 MAX ATC, FAA Order 8110.4C was in effect. Paragraph 2-3(d) of this order stated in part, "All TC applicants are required to submit a certification plan to the FAA and to keep it current throughout the project." The plan should be submitted early in the project and updated throughout the project." An NTSB review of this order found that it listed several key items that an applicant should include in its project certification plan. Some of the key items are the following:

- General information including applicant identification, application date, model designation, and so forth.

- A description of the proposed design or design change including sketches and schematics.

- The proposed certification basis including applicable regulation paragraphs and subparagraphs with amendment levels, exemptions, ELOS findings, and special conditions.

- A description of how compliance will be shown (ground test, flight test, analysis, similarity, or other acceptable means of compliance). The description of the means of compliance should be sufficient to determine that all necessary FAA data will be collected, and all findings can be made.
- A list of documentation that will be submitted to show compliance with the applicable certification basis, and how the applicant will ensure that all showings have been made. This can be accomplished using a compliance checklist addressing each section of the regulations applicable to the product.
- A project schedule including major milestones, such as preliminary hazard analysis submittal dates, substantiating data submittal dates, conformity and testing completion dates, and expected date of final certification.
- Identification of all designated manufacturing inspection representatives (DMIR), designated airworthiness representatives (DAR), and organizational designated airworthiness representatives (ODAR) intended for use, their authorized function codes, and their proposed inspection activities.
- For certification, the Certification Plan should list ARs/UMs and propose whether ODA be delegated to make compliance findings on behalf of the FAA.

H.1.2 Certification Plans - MCAS

Two Boeing certification plans (CP) address MCAS:
1. CP13471 Flight Controls – Primary, Elevator and Stabilizer Control, and

2. CP13474 Flight Controls – Autoflight (EDFCS/FCC) & Autothrottle.

Boeing was responsible for developing and updating these certification plans, submitting the plans to the BASOO for acceptance[43], and keeping the plans current throughout the design, development and certification phases of the 737 MAX project. An NTSB review of these two plans was conducted, and the findings are described below.

H.1.2.1 Certification Plan 13471 - Primary, Elevator and Stabilizer Control

CP13471, Revision AH, dated February 16, 2017, was reviewed by the NTSB to determine the methods (i.e., design test, analysis, inspection, etc.)

and approach Boeing used to demonstrate compliance to the applicable FARs. This version was the last revision before the Boeing 737-8 (MAX) amended type certificate was issued.

CP13471 detailed the activities necessary for the amended type certification of the flight controls aspects of the 737-8 Elevator and Stabilizer Control System changes. CP13471 indicated that the 737-8 will employ previously FAA-accepted methods of compliance which utilized industry standard analysis methods as well as Boeing standard analysis methods, tools and test procedures. Compliance will be demonstrated through analysis, qualification test, flight test and safety assessment using standard Boeing tools, methods and procedures. Testing to be completed under this certification plan includes Elevator Feel Computer qualification testing and Flight Testing for intended function for any new or modified systems.

rtification plan (CP13471), began with Boeing's initial submission of CP13471, labeled "NEW", to the FAA for review in March 2014. On March 29, 2016, Boeing received the FAA's acceptance of CP13471, Revision AA and the FAA indicated to Boeing that the implementation of their proposed certification activities could proceed. According to the delegation section of the plan, as of November 14, 2013, this certification plan was retained by the FAA and they would make a decision of delegation based on review of the certification plan.

According to CP13471, one of the changes to the Stabilizer Trim Control system from the baseline 737-800 (NG) was the incorporation of the MCAS. Implementation of this new function required two new analog discrete signals, generated by the FCCs, to be sent to components within the stabilizer system. One discrete will override the control column cut-out switches located in the First Officer's Column Switching Module in the "pull" direction when MCAS is operating to prevent the stabilizer command from cutting out during the pilot maneuver. The second discrete overrides the flap position input to enable the higher stabilizer trim motor (STM) operating speed with flaps retracted when MCAS is operating.

CP13471 indicated that certification of the MCAS implementation and function will be addressed in certification plan (CP13474), "737-8 Amended Type Certificate – Flight Controls – Autoflight (EDFCS/FCC)."

H.1.2.1.1 Cross Reference to Certification Plans

The stabilizer CP contained a section titled "Cross-Referenced Certification Plans" which detailed certification plans associated with this certification plan. As previously indicated, MCAS compliance information was contained in two certification plans; the Stabilizer CP (13471) and the EDFCS CP (13474). A review of the cross-reference section contained within the Stabilizer CP confirmed that it did reference the CP titled "Flight Controls –Autoflight EDFCS/FCC"; it also indicated that the EDFCS CP proposed a means to certify the 737-8 Autoflight Changes and specifically addresses the software changes required to implement revised Yaw Damper gains.

H.1.2.1.2 Functional Hazard Assessment (FHA)

CP13471, Revision AH, contained a section titled "Functional Hazard Analysis/System Safety Assessment Summary." According to the FHA, methods for assessing Functional Hazards included Pilot Simulation, Desktop Analysis, and Engineering Judgment. A select number of failure conditions will be flown for certification based on their probabilities and aircraft level effects on handling qualities. Failures that are extremely improbable or failures that were deemed Minor will not be flown. Complete system descriptions, hazard assessments and system safety analyses are referenced in deliverable #9 (Stabilizer System Safety Assessment). The functional hazard assessment identified and classified, pursuant to the guidance in AC 25.1309-1A, hazards associated with MCAS as noted below[44]:

- Catastrophic:

No catastrophic hazards were identified for MCAS

- Hazardous:
- 1.Uncommanded MCAS function operation until pilot recognition and reaction.
- 2.Uncommanded MCAS function operation to maximum authority.
- 3.Uncommanded MCAS function operation equivalent to 3 second mistrim.
- • Major/Minor:

No major or minor hazards were identified for MCAS

The NTSB notes that the FHA classification of uncommanded MCAS operation varied depending on whether the aircraft was in the normal or operational flight envelope. CP 13471 lists only the most severe of the two, which in the case of the operational flight envelope is "Hazardous".

H.1.2.1.3 Delegation of Deliverables

CP13471 proposed delegation of all Flight Controls Primary & Secondary compliance findings. On April 14, 2015, the FAA approved the delegation of several deliverables; however, they indicated that the deliverable titled "737 Stabilizer System Description and Safety Analysis" (SSA) would be retained by the FAA and will not be proposed for delegation. In November 2016, Boeing submitted the 737 Stabilizer System Description and Safety Analysis (SSA), revision F, to the FAA for acceptance. "In December 2016, the FAA's response to Boeing was to "accept" the submittal and with notation "delegated SSA approval to ODA."
Retention and delegation are accomplished with respect to compliance deliverables not to specific functions i.e., MCAS itself would not be delegated to the ODA.

- Consistent with the FAA authorization, the FAA have discretionary authority as to what is reviewed, whether submitted directly to the FAA for review and approval by an applicant or submitted by a designee or ODA recommending approval.

- When delegating at the end of a program, there has been some level of FAA involvement and the delegation confirms that the designee should make the final approval.

- In all cases, delegation is not accomplished by a single individual but follows a structured review process.

H.1.2.1.4 Method of Compliance (MOC)

CP13471 indicated that a Stabilizer System Safety Analysis (SSA) will show that the Stabilizer System including both the changed and unchanged designs meet the reliability, integrity and safety requirements for the 737-8 aircraft. The SSA will include a Failure Modes and Effects Analysis, Functional Hazard Assessment and Fault Tree Analysis.

H.1.2.1.5 Deliverable Matrix

CP13471 contained a section titled "Deliverable Matrix" which provides a description of the deliverable[44], the method of compliance, FAA requirements, and Approver. The NTSB's review of CP13471 found the deliverable related to MCAS was the 737 Stabilizer System Description and Safety Analysis (SSA). This document provides the complete details of the installation, interfaces, design features, control and operation of the stabilizer control system. The Stabilizer SSA also contains all top-level failure conditions or safety issues, the failure effect category according to each condition and the appropriate supporting analysis identified during the functional hazard assessment.

H.1.2.2 Certification Plan – Autoflight (EDFCS/FCC) & Autothrottle

EDFCS consists of two Flight Control Computers (FCCs), one Mode Control Panel (MCP), and one Integrated Flight Systems Accessory Unit (IFSAU). The EDFCS provides Autopilot, Flight Director, Mach Trim, Speed Trim, Altitude Alert, and Autothrottle functions.

The development of EDFCS certification plan (CP13474) began with Boeing's initial submission of CP13474, revision "NEW", to the FAA for review in March 2014. On June 2, 2015, Boeing received the FAA's acceptance of CP13474, Revision F and the FAA indicated to Boeing that the implementation of their proposed certification activities could proceed. CP13474, revision U, dated February **28, 2017, was reviewed by the NTSB to determine the methods** (i.e.,design test, analysis, inspection, etc.) and approach Boeing used to demonstrate compliance to the applicable FARs. This version was the last revision before the 737-8 amended type certificate was issued.

A review of CP13474 found that the changes to the EDFCS for the 737-8, as compared to the baseline 737-800, were limited to the Flight Control Computer (FCC) software only. CP13474 indicated that the FCC Operational Program Software (OPS) will be revised to add the MCAS function.

H.1.2.2.1 Cross Reference to Certification Plans

The EDFCS CP contained a section titled "Cross-Referenced Certification Plans" which detailed certification plans associated with this certification plan. A review of the cross-reference section contained within CP13474 confirmed that it did reference the CP13471 titled "737-8 Amended Type Certificate – Flight Controls – Primary, Elevator and stabilizer Control; it also indicated that CP13471 proposed a means to certify the 737-8 Elevator and Stabilizer Control system changes, including testing and analysis for the Maneuvering Characteristics Augmentation System (MCAS).

H.1.2.2.2 Compliance Matrix

A review of CP13474 found that it contained a compliance matrix for FAA advisory circular (AC) 25.1329 Approval of Flight Guidance Systems. The compliance matrix included a table showing the proposed compliance statement and the deliverables. According to the table, a System Safety Analysis (SSA), will provide an assessment of the EDFCS as part of an integrated system to the extent that such interactions affect the top-level hazards derived from the FHA. An aircraft-level assessment of multiple system failure combinations will be address by the single and multiple failure analysis conducted by Airplane Safety Engineering Organization.

The SSA will be performed in accordance with 14 CFR 25.1309 and AC 25.1309-1A. Common mode/cause or cascading failures will be evaluated. The existing EDFCS SSA will be updated and revised as required for the 737-8.

The SSA will provide an assessment of the EDFCS hazards in the summary FHA and all possible failure modes in the EDFCS and its interfacing systems. This assessment will include consideration of interactions with other systems and the effects of failure combinations of sensors and systems on flight crew workload, aircraft structural integrity, and occupant safety in accordance with AC 25.1309-1A. The SSA will be validated through analysis, lab test, simulation and flight test as appropriate. The SSA will provide documentation of the validation methods.

H.1.2.2.3 Functional Hazard Assessment

CP13474 contained a section titled "Functional Hazard Analysis/System Safety Assessment Summary." According to this FHA, the EDFCS Functional Hazard Assessment for the 737-8 will be based on the FHA for the 737NG as documented in the document titled "Enhanced Digital Flight Control System, Autothrottle, and Yaw Damper Safety Analysis, Model 737-600/700/800/900." CP13474 indicated that the FHA was to be updated to address any functional hazards associated with the addition of the Maneuvering Characteristics Augmentation System **(MCAS), and other system changes.**

H.1.2.2.4 Software/Airborne Electronic hardware Considerations

A review of CP13474 found that it contained a table describing a discussion on the software used in the FCC's. According to the discussion, the software will be developed by Rockwell Collins in Cedar Rapids, Iowa. Rockwell Collins will create a Plan for Software Aspects of Certification

(PSAC) based on the guidance of FAA Advisory Circular 20-115B, RTCA/DO-178B, the RTCA/DO-178B errata in RTCA/DO-248B and FAA Order 8110.49 Change 1. The PSAC will contain the preliminary software change impact analysis and will be available following certification plan approval. It is proposed to have a Rockwell Collins Software OBAR[46] to make the compliance findings.

H.1.2.2.5 Delegation Discussion

CP13474 indicated that approval of the EDFCS System Safety Analysis would be retained by the FAA and would not be proposed to be delegated to the Boeing ODA. The FAA retained approval of the SSA until revision K, submitted in January 2017. At that time, the FAA stamped the revision as "rejected" due to the need to correct some information and simultaneously delegated approval of the SSA once the final edits were complete.

H.1.2.2.6 Method of Compliance

In the Method of Compliance section of the CP, Boeing proposed that the System Authorized Representative (AR) would review the applicable deliverables in this certification plan to verify the compliance and its proper documentation.

An EDFCS System Safety Analysis will show that the system design meets the reliability, integrity, and safety requirements for the 737-8 aircraft. The document will include a Failure Modes and Effects Analysis, Functional Hazard Assessment, and the Fault Tree Analysis to demonstrate compliance to the applicable regulations, FAA AC 25.1309-1A, FAA Issue Paper S-1 and EASA CRI D-09.

The software will be verified by design and process reviews per the standards of DO-178B appropriate to the design assurance level. DO-178B is an FAA approved means of compliance for software per AC20-115B. The software will be developed by Rockwell Collins Inc in Iowa. The Software Accomplishment Summary (SAS) will be a summary of all design development and verification activities defined in the PSAC that provides the data to substantiate that the objectives of RTCA DO-178B for the appropriate design software level have been met. The Systems AR approval/recommend approval is limited to the integration of system requirements/functionality to the software.

H.1.2.2.7 Deliverable Matrix

CP13474 contained a section titled "Deliverable Matrix" which provides a description of the deliverables, the method of compliance, FAA requirements, and Approver. The NTSB's review of CP13474 found the deliverables (compliance documents) directly related to MCAS:

- Software Accomplishment Summary:

The Software Accomplishment Summary for the Flight Control Computer (FCC-730) shows the compliance of the Flight Control Computer software development and verification to the Plan for Software Aspects of Certification for the FCC-730. Delegation of this deliverable was granted via an FAA response on 4/18/2016.

- Final Enhanced Digital Flight Control System Safety Analysis:

This document presents the system safety assessment for the 737-8 Enhanced Digital Flight Control System.

- Final Enhanced Digital Flight Control System Description Document:

The Enhanced Digital Flight Control System Description document provides a description of the 737-8 EDFCS, including a description of all EDFCS components, functions, maintenance and ground operations, crew interfaces, and aircraft interfaces.

H.2 Safety Assessments

Safety assessments are a primary means of showing compliance for systems to FAR 25.1309. Safety assessments proceed in a stepwise, data-driven fashion, analogous to the system development process described above. Starting with aircraft functions, functional hazard assessments are performed to identify the failure conditions associated with each function. Systems functional hazard analyses are performed for system level functions. Preliminary safety assessments are performed as the system is developed adding more specific design and implementation detail to address specific hazards. The bottom-up verification by safety analysis starts with an analysis of the components of a system to ensure single failures do not result in significant effects. Combinations of failures are logically combined to develop probability of a failure and checked to ensure they are commensurate with the criticality of the failure condition. Thus, the final definition and characterization of a safety-critical system is verified by the result of the analyses conducted during a safety assessment.

As previously stated, certification plans CP13471 & CP13474 each indicated that a system safety analysis (SSA) would be a method of compliance and a deliverable to their respective certification plan. An NTSB review of CP13471, revealed that a Stabilizer SSA will show that the changed and unchanged designs of the Stabilizer System meet the reliability, integrity and safety requirements for the 737-8 aircraft. The review also showed that for CP13474, an EDFCS SSA will show that the system design meets the reliability, integrity, and safety requirements for the 737-8 aircraft. The SSA documents will include a Failure Modes and Effects Analysis, Functional Hazard Assessment, and the Fault Tree Analysis to demonstrate compliance to the applicable regulations, FAA AC 25.1309-1A, FAA Issue Paper S-1 and EASA CRI D-09. Because the 737 MAX Air Data Inertial Reference System SSA, discussed the Angle-of-Attack (AOA) sensors and its failure modes, the NTSB also performed a review of this SSA. The following sections describe these three SSA's in greater detail.

H.2.1 Air Data Inertial Reference System (SSA)

Boeing's 737 MAX Air Data Inertial Reference System SSA, dated August 12, 2016, Revision New, was a deliverable to Certification Plan CP13486 titled, "737-MAX Air Data Inertial Reference System Certification Plan." The NTSB performed a review of this SSA and documented information that pertained to Angle-of-Attack (AOA) sensors.

A description of the Air Data Inertial Reference System (ADIRS) was provided in the SSA, it indicated that the Air Data Inertial Reference Unit (ADIRU) consisted of an Air Data Reference partition and an Inertial Reference partition packaged into a single unit. The two partitions are physically separate and operate as separate functions including independent inputs and outputs for each.

The SSA indicated that the function of the Air Data Module (ADM) is to sense the aircraft pitot and static pressures external to the aircraft and convert them to a digital electrical signal. These pressures, in conjunction with the Total Air Temperature (TAT) and the aircraft AOA are used to calculate the basic air data information. The ADIRU then transmits the data (several parameters including indicated angle of attack), via ARINC 429 busses, to other systems for display to the flight crew, use in flight control functions (including MCAS), and other aircraft system functions.

With regards to AOA, the SSA provided a description of the AOA sensor that stated the following; two independent sensors are used to provide AOA data to the air data partition of the ADIRU". It also indicated that

the two vanes are located on each side of the aircraft fuselage and measure the aircraft AOA relative to the local air mass. The output of the AOA internal electrical transducers (resolvers) is input directly into the ADIRU, which then outputs an indicated AOA signal to other systems.
NTSB Note:

Each AOA sensor has two resolvers within it, one of which is connected to the associated ADIRU. The other resolver in each AOA sensor is connected to a stall management yaw damper (SMYD) computer.
The SSA indicates that the altitude and airspeed functions within the ADIRU include a correction factor for Static Source Error (SSEC[47]). This is a compensation for pressure errors caused by the airframe aerodynamic effects on the static port which predictably vary with AOA and Mach number.

The SSA contains a section titled "Angle of Attack Failure," which states:
"The Angle of Attack Vane senses the alpha angle of the airplane. The Static Source Error Correction (SSEC) is calculated as a function of indicated Mach and AOA. Therefore, all parameters which are based on corrected static pressure are impacted if the AOA vane fails. Also, since the AOA vane has only two resolver output circuits, the AOA is also provided as an output to other systems. The following parameters (shown in table 4) will be output as No Computed Data (NCD)":

The "Angle of Attack Failure" section of the SSA includes only AOA resolver circuit failures (open circuit, high impedance, etc.) that can be detected by the associated computer (ADIRU or SMYD). The SSA does not discuss the category of AOA sensor failures not related to the electrical circuitry that could provide misleading (erroneous) data to the ADIRU (e.g. a frozen or seized vane with limited or no motion, or a bent or broken vane resulting in angular offset). As demonstrated by the Lion Air event, erroneous input from the AOA sensor affects the calculation of the SSEC and thus all parameters based on the measurement of static pressure (including airspeed and altitude). However, this failure will not result in the parameters being output as No Computed Data (NCD). Instead, AOA values are transmitted as "valid" to user systems, because the ADIRU does not detect these faults.

The SSA contained a table summarizing the results of a failure analysis by functional group. The results were provided in terms of loss of function (detected failures) and misleading data (undetected failures) for each primary group. For the misleading data rates, the fault trees were reviewed to determine which components of the system could contribute to misleading information. It was determined that the ADIRU, air data module (ADM), pitot probe heat and AOA vane (and heat) have potential undetected failure modes that may result in undetected, and misleading data.

The SSA concludes that the ADIRS is a primary sensor on the aircraft and it supports many aircraft functions. By itself, the ADIRS is not required to satisfy specific functional failure rates for either independent side or as a system. However, as an input to other aircraft systems, the ADIRS must provide functional failure rates to allow those systems to satisfy the aircraft functional failure rate requirements.

The SSA also contained a section that summarized the functional failure conditions. This section indicated that some aircraft display requirements drive the ADIRS requirements.

The SSA contained a section titled "Fault Tree Documentation" that developed fault trees for the events that were identified as hazardous or catastrophic. An NTSB review of these fault trees was conducted to determine if and how Boeing considered the effects of a single AOA sensor providing a "loss of data" or "misleading data (erroneous data)" to aircraft systems.

The review found that Boeing did consider the effects of a single AOA sensor providing a "loss of data" within a fault tree with the "Top Event" titled "loss of AOA data for both sides". Or basically, there is a loss of AOA data from both the left and right side ADIRU's. The fault tree showed that there were two failure conditions contributing to this "Top Event": 1) Loss of number 1 AOA and 2) Loss of number 2 AOA. For each of these failure conditions, one of the contributing factors was "No AOA output to the ADIRU" which could result from either of the following basic events:

- An AOA Vane failure, or
- Loss of power

The review also found that Boeing considered the effects of a single AOA sensor providing "erroneous data" within the lower branches of a fault tree with the "Top Event" titled "Misleading Air Data from the Left and Right ADIRU – Airspeed / Altitude." The fault tree showed there were two failure conditions that contributed to this top event:

- Misleading Air Data from the Left ADIRU, and
- Misleading Air Data from the Right ADIRU

The two failures are symmetric, the left is considered here. According to the fault tree, there are four failure conditions that contribute to the "Misleading Air Data from the Left ADIRU" hazard. One of these conditions was titled "Erroneous AOA-L data from the Captain's side"; the other three were not related to the AOA sensor itself. The fault tree showed the following two ways (or failure conditions) that could lead to "Erroneous AOA-L data from the Captain's side".

- *"failure of AOA-L vane / Annunciation"*

- *"incorrect AOA output from the ADIRU-L output."*

For the *"failure of AOA-R vane / Annunciation"*, the fault tree showed that this event could occur by the combined (ANDed) result of the following two failure conditions:

- "Loss of AOA-L Heat Annunciation"

- "Erroneous AOA-L Sensor"

In 2019, Boeing advised the NTSB of an error in this fault tree in that the above two conditions should not have been combined with an AND gate. In a June 28, 2019 revision to the SSA, "Erroneous AOA-L data from Captain's side" is revised to show three separate conditions combined with on OR gate, meaning any one by itself could result in erroneous AOA data:

- Erroneous AOA-L Sensor

- Incorrect AOA output from ADIRU-L output

- Loss of Power to AOA-L Heater

In both the original and revised fault tree, the top event "Misleading Air Data from L & R ADIRU – Airspeed/Altitude" showed that it met the requirement to be extremely improbable.

H.2.2 Stabilizer Trim Control System Safety Analysis (SSA)

Boeing's 737 NG/MAX Stabilizer Trim Control SSA was a deliverable to Certification Plan CP13471 titled, "Flight Controls – Primary, Elevator and Stabilizer Control." The SSA was originally developed to provide a safety analysis showing compliance with the certification agency requirements for the 737-6/7/8/900 (737 NG) family of aircraft. The safety analysis included a description of the Stabilizer Trim Control

System, tables for certification and Means of Compliance, a Functional Hazard Assessment (FHA) summary, Failure modes and effects analysis, and fault tree analysis (FTA).

The NTSB performed a review of Revision H, dated November 28, 2017 of this analysis and documented information that pertained to the incorporation of MCAS in the following paragraphs. Revision F of the SSA document, dated September 7, 2016, incorporated a new Appendix G, which contained the safety analysis for the 737 MAX family of aircraft. Appendix G was added to document the safe operation of the 737 MAX stabilizer trim system and to show compliance with

certification agency requirements. Included within this Appendix are sections that provide a system description of MCAS, a Functional Hazard Assessment (FHA) summary which identified the severity of potential hazards to the aircraft due to the implementation of the 737 MAX stabilizer trim system changes, and fault tree analysis (FTA) documentation showing that for identified top failure scenarios, the probability of occurrence is less than extremely improbable (1E-9). Then the summary converse statement can be made that for the 737 MAX stabilizer trim system, the aircraft is capable of continued safe flight and landing without requiring exceptional pilot skill or strength, following any combination of failures not extremely improbable.

An NTSB review of Appendix "G" found that the introductory section of SSA had not been updated to reflect the March 2016 MCAS maximum authority changes. The introductory section indicated that MCAS was added on the 737 MAX to address potentially unacceptable nose-up pitching moment at high angles of attack at high airspeeds; there was no mention that MCAS had been revised to improve flaps up, low Mach tall characteristics and identification. Additionally, the functional hazard assessment summary table contained within the Appendix still reflected a pre-March 2016 MCAS maximum authority limit of 0.6 degrees.

However, an NTSB review of Boeing internal documents confirmed that the FHAs had in fact been reassessed each time that the MCAS requirements were changed, including the change in authority limit from 0.6 to 2.5 degrees. In all cases, the reassessment found that the FHA categories had not changed.

In responses to an NTSB request, the FAA provided the following response for how they became aware of the March 2016 changes to MCAS (i.e. improved flaps up, low Mach stall characteristics and identification) and if the group within the BASOO who was responsible for approving the Stabilizer system safety assessment (SSA) were aware of the change. The FAA indicated the following:

In a July 2016 briefing[48], Boeing provided the FAA with a presentation on stall characteristics and configuration changes. The purpose of this briefing was to discuss company test results prior to entering into certification testing. At this briefing, Boeing discussed some of the physical aerodynamic devices (relocation of stall strip, vortex generators (VG) configurations, etc.) they used to improve the stall characteristics with only limited success. During the briefing Boeing discussed their intent to expand the MCAS function to activate at lower Mach speeds. The actual amount of authority was not defined at that time as Boeing was still conducting testing to tune and validate the system. FAA well understood that greater MCAS authority would likely be necessary to cover the lower speed region. In July 2016, Boeing provided a similar presentation to the FAA with additional company test results. Based on those results, Boeing finalized the MCAS design tables and submitted their revised certification plan in September 2016. Numerous validation meetings were held in the Fall of 2016 (CAAC, TCCA, EASA), supported primarily by FAA flight test and the policy office. In those meetings, the maximum MCAS authority of 2.5 deg in the low speed region was specifically covered. The FAA also indicated that their focus on the SSA's was mainly around other system changes and not MCAS and therefore from a flight controls / system safety perspective their team does not have recollection of specific discussions associated with Boeing regarding the MCAS changes.

H.2.2.1 Functional Hazard Assessment

The Functional Hazard Assessment section of Appendix "G" summarized the FHA that was performed as part of the 737 MAX Stabilizer Trim Control System Safety Analysis, and addressed each system function and the result of loss of availability or loss of integrity of that function. The analysis considered all phases of flight for both the Normal and Operating flight envelopes[49], interfacing systems, and established the effect category for each failure condition. Hazard assessments were determined in consideration of the impact to crew workload for the maximum flight time and longest diversion time (where a diversion is required). An NTSB review of the FHA found that it identified and classified, pursuant to the guidance in AC 25.1309-1A, six hazards related to MCAS.

H.2.2.2 Fault Tree Analysis

Appendix "G" contained a section titled "Fault Tree Analysis (FTA)" that presented the fault trees that were developed as part of the Stabilizer Trim Control System safety analysis. According to the analysis, FTA is a tool used to quantitatively determine the numerical probability of a certain combination of events. The failure conditions defined by the FHA provide the basis for the top-level events analyzed by the FTA to demonstrate compliance with 14 CFR 25.671(c)(2), (c)(3), and 25.1309(b)(1).

Boeing indicated that fault tree analyses were only performed on the FHA events that were determined to be either Catastrophic or Hazardous, which is consistent with the guidance in SAE ARP 4761. As described above, unintended MCAS activation was shown to be Major in the normal flight envelope and Hazardous in the operational flight envelope. FAA Advisory Circular 25-7C Appendix 5 lists the probability of being outside the normal flight envelope as 1E-3. Therefore, a condition that meets the integrity requirements for a Major within the normal flight envelope also meets the Hazardous integrity requirements for the operational flight envelope.

Therefore, unintended MCAS operational FHA events were not evaluated in the fault tree analysis as they were assessed as Major in the normal flight envelope; Boeing indicated that is consistent with FAA regulations and the Boeing process.

Although the failure conditions (such as a single AOA failure) that could result in an unintended MCAS operation were not evaluated as part of Boeing's Stabilizer System fault tree analysis, an NTSB review of their analysis found that Boeing had modified (updated) their original (737 NG) catastrophic fault trees to account for MCAS engage discrete failures which could contribute to a loss of the control column cutout function.

H.2.3 EDFCS Auto-throttle, and Yaw Damper System Safety Analysis (SSA):

Boeing's 737 NG/MAX Enhanced Digital Flight Control System, Autothrottle and Yaw Damper Safety Analysis (SSA) was a deliverable to Certification Plan CP13474. The NTSB performed a review of the EDFCS SSA, Revision M, dated January 24, 2018. The review found that Boeing had added an appendix (Appendix E) to the original SSA to document the information specific to the 737 MAX. Relevant to MCAS, the appendix included the following sections: compliance summary, Summary of system changes, MCAS description, and a fault tree analysis. According to the "compliance summary" section, the EDFCS changes incorporated in the 737 MAX were evaluated for impact to the baseline safety analysis provided in the main body of the SSA (analysis for the 737-300/400/500/600/700/800/900 aircraft). The "Summary of system changes" section indicated that the EDFCS architecture in the 737 MAX is the same as in the 737 NG and the changes to the system are limited to the software resident in the FCCs. The software changes support the addition of new EDFCS functionality for the 737 MAX, including the Maneuvering Characteristics Augmentation System (MCAS) and other systems. To incorporate MCAS, the following was required: two new MCAS related FCC discrete outputs; modifications to the Column Switching Module and Stabilizer Trim Motor interface wiring.

MCAS would be active during manual flight only and would drive the stabilizer in flaps-up, high angle of attack conditions to improve pitch-up handling characteristics. The FCC software revisions include the following:

•   Logic to prioritize and command the stabilizer trim motor for MCAS operations using the active Speed Trim channel.

•   Output of MCAS Engage discretes from the FCC in command, to set the high stabilizer trim motor rate and inhibit the column cut-out function of the Column Switching Module in the aft direction.

H.2.3.1 Baseline Analysis – Background

According to the SSA, the Digital Flight Control System (DFCS) which provided the autopilot function on 737-300/400/500 and early 737-600/700/800/900 (737NG) aircraft was replaced by an upgraded version developed by a different supplier. This upgraded version, known as the Enhanced Digital Flight Control System (EDFCS), was introduced in 2004 and used in all 737NG aircraft delivered since then as well as all 737MAX aircraft. The primary purpose of the SSA was to document the systems compliance to the safety requirements of 25.671, 25.672, and 25.1309. A Functional Hazard Assessment defines the hazards of interest. The system FMEA ensures that no single failure will cause a Catastrophic event. The fault trees examine the probability of combinations of faults which could contribute to a hazard in manual flight, autothrottle on, single channel autopilot, and dual channel fail-passive or fail-operational autoland operation.

The general safety analysis process provided in ARP 4761, "Guidelines and Methods for Conducting the Safety Assessment Process on Civil Airborne Systems & Equipment" was used as a general guide in performing the analysis. Each system component or interfacing system was investigated to determine if any failure modes exist which could contribute to one of the functional hazards or cause a loss of fail-passive or fail-operative capability during autoland. Systems such as hydraulic power, electrical power, computers and sensors, and electrical wiring were examined to assure that adequate isolation was provided between redundant systems. System interlocks, monitoring, and warning systems were studied to ensure that these systems would protect against significant failures. Unique functions were also analyzed that could violate the brickwall architectural approach or monitoring independence if improperly implemented or applied (cross-channel communications, equalization, synchronization, initialization processes, localizer averaging, monitor testing and isolation, etc.)

The SSA included a detailed Functional Hazard Assessment, a Failure Modes and Effects Analysis (FMEA) that presented an analysis of failure modes particularly relevant to each system operation, a section that provides the results of simulated and flight test worst case failure evaluations and an assessment of the effects of potential pilot errors related to the operator interface. The system definition, functional hazard assessment, and the failure modes and effects analyses were the sources for the fault tree analyses. The analysis also included fault tree assessments for the top-level events defined by the Functional Hazard Assessment.

H.2.3.2 Requirements

The SSA contained a table describing the FAA requirements and the method of compliance. One of the requirements was 14 CFR 25.672 "Stability augmentation and automatic and power-operated systems", which states: If the functioning of stability augmentation or other automatic or power-operated systems is necessary to show compliance with the flight characteristics requirements of this part, such systems must comply with §25.671 and the following:

(a) A warning which is clearly distinguishable to the pilot under expected flight conditions without requiring his attention must be provided for any failure in the stability augmentation system or in any other automatic or power-operated system which could result in an unsafe condition if the pilot were not aware of the failure. Warning systems must not activate the control systems.

(b) The design of the stability augmentation system or of any other automatic or power-operated system must permit initial counteraction of failures of the type specified in §25.671(c) without requiring exceptional pilot skill or strength, by either the deactivation of the system, or a failed portion thereof, or by overriding the failure by movement of the flight controls in the normal sense.

(c) It must be shown that after any single failure of the stability augmentation system or any other automatic or power-operated system—

(1) The aircraft is safely controllable when the failure or malfunction occurs at any speed or altitude within the approved operating limitations that is critical for the type of failure being considered;

(2) The controllability and maneuverability requirements of this part are met within a practical operational flight envelope (for example, speed, altitude, normal acceleration, and aircraft configurations) which is described in the Aircraft Flight Manual; and

(3) The trim, stability, and stall characteristics are not impaired below a level needed to permit continued safe flight and landing.

According to the SSA, Speed Trim, MCAS, Mach Trim and Yaw Damping are augmentation functions covered by Section 25.672. These augmentation functions comply with 25.671 and, following a system failure, can be deactivated or overridden by the pilot without exceptional skill or strength. In addition, the aircraft is capable of continued safe flight and landing following any failures not extremely improbable.

A review of functional hazard assessment found that it addressed each system function and the result of loss of function or erroneous operation. The analysis considered phases of flight, flight envelope, interfacing systems, and established effect categories for the failure conditions. According to the SSA, the FHA analysis was reviewed by all affected organizations including: Flight Controls, Aero/S&C, Flight Deck, Pilots, Reliability, Safety, and Structures. Performance analysis or simulation were accomplished as needed to help define the hazards or criticality. Lab and flight test conditions for validation of the assignment of criticality of specific hazards were defined as well. Each FHA event was closed by reference to a specific analysis, design feature, or test condition. However, because MCAS only operates with the autopilot off, one hazard contained within the assessment was relevant. This hazard is: "Autoflight Malfunction at Low Altitude Which Results in Unsafe Flight Path in an A/P OFF, Single Channel, or Fail-Passive Configuration (FHA 1).

According to the fault tree analysis section of Appendix E, the original (Baseline 737 NG) fault trees contained in the EDFCS SSA document were assessed for applicability to the 737 MAX and were modified as needed to account for functional changes specific to the 737 MAX configuration.

With regards to MCAS, the SSA indicated that the inclusion of the new MCAS function creates new failure modes affecting the probability of runaway stabilizer trim which cannot be arrested by the column cutout switches. As previously described, the MCAS function normally activates only during manual flight, and operates by trimming the horizontal stabilizer in the nose-down direction while the aircraft is executing a high AOA maneuver. The column cutout switch mechanism normally inhibits automatic nose-down automatic trim in the presence of aft column inputs applied by the flight crew. The MCAS implementation therefore required a new relay that provided a bypass of the column cutout switches when the MCAS Engage discrete is asserted by either FCC. Any erroneous activation of the MCAS Engage output will energize the bypass relay and prevent aft column inputs from interrupting nose-down automatic trim commands.

To account for this hazard[50], Boeing modified the fault tree for the failure conditions titled "Erroneous Runaway/oscillatory stab output un-arrested by column cutout". This failure condition was one of eight conditions that contributed to a higher-level failure condition titled "Autopilot Malfunction in the Pitch Axis at Low Altitude." And, this failure condition is one of four conditions that contributes to the Top-Level event titled "Autoflight malfunction at low altitude which results in an unsafe flight path in an autopilot OFF, single channel or fail passive configuration," This Top-Level event was identified as a catastrophic hazard as part of Boeing's EDFCS functional hazard assessment.

An NTSB review of the modifications incorporated into the fault tree titled "Erroneous Runaway/oscillatory stab output un-arrested by column cutout" revealed that the following two failure conditions "AND'ed" together resulted in the hazard.

- Column Trim Cutout Fails to Interrupt Stab Motion
- Undetected stabilizer trim runaway

For the "Column Trim Cutout Fails to Interrupt Stab Motion" hazard, the fault tree identified two potential failure conditions (OR'ed together) that could result in the hazard. One of the failure conditions "FCC-730 produces undetected erroneous MCAS or Flaps Up/Dn discrete" is where the fault tree begins to address the erroneous activation of the MCAS Engage outputs and is also where Boeing introduced SCD requirement "3.1.1.5.3.1.1-A" which set up upper limit on "The probability of the FCC producing an erroneous flaps up/down discrete output or an erroneous MCAS Engage discrete output without detection." For this event, the fault tree showed the requirement was satisfied.

Tracing the failure conditions that could lead to the hazard identified by the SCD led to the event titled *"input failures cause FCC to produce an undetectable erroneous MCAS engage discrete"* The probability for this event was <1E-9"

H.3 Single and Multiple Failure Analysis

the deliverables required by a certification plan, Boeing performed a S&MF analysis to help validate system functional hazard assessments (FHAs), design assurance level (DAL) assertions, and extended operations (ETOPS). For the 737 MAX, their analysis is contained within a document titled "Single and Multiple Failure Accomplishment Summary 737 MAX Program," revision New, which was released on January 19, 2016. According to Boeing, the S&MF analysis was started internally in 2014.
The intent of this analysis was to provide a structured methodology to analyze failures of key integration components and functions to determine if aircraft, flight crew, and occupant impacts are as expected and acceptable. The analysis includes intersystem failures and their cascading effects, flight deck indications, and pilot procedures. It was performed on failures originating in one or more systems with multi-system effects that are not understood without an aircraft-level review.

An NTSB review of the S&MF Accomplishment Summary document revealed that the S&MF analysis process was led by a Boeing Systems Engineering team along with design engineers, system subject matter experts (SMEs), systems engineers, Safety, Crew Operations, and pilots. Other representatives also participated as appropriate (e.g., Aerodynamics, authorized representatives, etc.). Until completed, the S&MF analysis process was iterative, and the analysis was updated if significant change affected key systems. Once completed prior to flight test, there was no requirement to redo the S&MF analysis for subsequent design changes

The S&MF analysis consisted of individual cases, each of which may contain one or more failures. Analysis cases were identified by members of the team using S&MF documents from previous programs, aircraft architecture descriptions and areas of change, schematics, system safety analyses and other information. The Boeing team defined each S&MF analysis case according to guidance material contained within a Boeing manual titled "Conducting Single and Multiple Failure Analyses." This guidance was used to help the multi-discipline team to choose the S&MF cases for the MAX program. Some cases were selected based on authorized representative (AR) requirements to show cases were acceptable and for specific conditions based on common cause failures. Some candidate cases were excluded for reasons such as: duplicate and/or mirror-image candidates, worst case candidates would often replace multiple less-severe cases, etc.

The Boeing team considered including "Erroneous AOA from a single source" as a case in the S&MF, but ultimately did not, identifying other multiple failures conditions that presented a more severe hazard to the aircraft. These conditions included "Erroneous L&R Air Data" and "Loss of one AOA followed by Erroneous AOA". These multiple failure cases were rated as catastrophic because they could result in all of the air data on the primary displays being misleading. Un-commanded MCAS was documented as a potential consequence of erroneous AOA, but was not identified as a factor contributing to the catastrophic rating in any of these. The acceptability rationale for these cases noted that these multiple failure events was beyond extremely improbable. The rationale also noted that while the failure event was catastrophic before flight crew recognition, training would support

H.3 Single and Multiple Failure Analysis

Although the single and multiple failure (S&MF) analysis was not one of the deliverables required by a certification plan, Boeing performed a S&MF analysis to help validate system functional hazard assessments (FHAs), design assurance level (DAL) assertions, and extended operations (ETOPS). For the 737 MAX, their analysis is contained within a document titled "Single and Multiple Failure Accomplishment Summary 737 MAX Program," revision New, which was released on January 19, 2016. According to Boeing, the S&MF analysis was started internally in 2014.

The intent of this analysis was to provide a structured methodology to analyze failures of key integration components and functions to determine if aircraft, flight crew, and occupant impacts are as expected and acceptable. The analysis includes intersystem failures and their cascading effects, flight deck indications, and pilot procedures. It was performed on failures originating in one or more systems with multi-system effects that are not understood without an aircraft-level review.

An NTSB review of the S&MF Accomplishment Summary document revealed that the S&MF analysis process was led by a Boeing Systems Engineering team along with design engineers, system subject matter experts (SMEs), systems engineers, Safety, Crew Operations, and pilots. Other representatives also participated as appropriate (e.g., Aerodynamics, authorized representatives, etc.). Until completed, the S&MF analysis process was iterative, and the analysis was updated if significant change affected key systems. Once completed prior to flight test, there was no requirement to redo the S&MF analysis for subsequent design changes

The S&MF analysis consisted of individual cases, each of which may contain one or more failures. Analysis cases were identified by members of the team using S&MF documents from previous programs, aircraft architecture descriptions and areas of change, schematics, system safety analyses and other information. The Boeing team defined each S&MF analysis case according to guidance material contained within a Boeing manual titled "Conducting Single and Multiple Failure Analyses." This guidance was used to help the multi-discipline team to choose the S&MF cases for the MAX program. Some cases were selected based on authorized representative (AR) requirements to show cases were acceptable and for specific conditions based on common cause failures. Some candidate cases were excluded for reasons such as: duplicate and/or mirror-image candidates, worst case candidates would often replace multiple less-severe cases, etc.

The Boeing team considered including "Erroneous AOA from a single source" as a case in the S&MF, but ultimately did not, identifying other multiple failures conditions that presented a more severe hazard to the aircraft. These conditions included "Erroneous L&R Air Data" and "Loss of one AOA followed by Erroneous AOA". These multiple failure cases were rated as catastrophic because they could result in all of the air data on the primary displays being misleading. Un-commanded MCAS was documented as a potential consequence of erroneous AOA, but was not identified as a factor contributing to the catastrophic rating in any of these. The acceptability rationale for these cases noted that these multiple failure events was beyond extremely improbable. The rationale also noted that while the failure event was catastrophic before flight crew recognition, training would support flight crew recognition and drive appropriate flight crew response to the flight deck effects (which, as noted above, included MCAS activation).

Boeing advised that after the accident, they reviewed how the case of single erroneous AOA would have been categorized if included in the original review. Boeing concluded that had the case of "Erroneous AOA from a single source" been included in the S&MF document, the same assumption about pilot response to un-commanded MCAS as used in the FHAs (which was based on regulatory guidance in AC 25-7C) would have been used, and it is unlikely that any design changes would have resulted from including this case in the S&MF analysis. As noted in section E.1, Boeing did conduct a similar, less formal analysis of the effects of erroneously high AOA on MCAS and concluded that no redesign was needed.

The S&MF Analysis was completed and published in January 2016. In March of that year, the MCAS authority was increased from 0.55 to 2.5 degrees. The NTSB notes that the S&MF analysis had been completed prior to the MCAS design change and was not re-visited as a result of the change[51].

I FLIGHT TEST GUIDANCE FOR CERTIFICATION OF TRANSPORT CATEGORY AIRPLANES

FAA advisory circular (AC) 25-7C, titled, "Flight Test Guide for Certification of Transport Category Airplanes," dated October 16, 2012, provides guidance for the flight test evaluation of transport category airplanes. AC 25-7C includes flight test methods and procedures to show compliance with the regulations contained in subpart B of Title 14, Code of Federal Regulations (14 CFR) part 25, which address aircraft performance and handling characteristics. Revision C to AC 25-7, was a complete revision to reduce the number of differences from the European Aviation Safety Agency's Flight Test Guide, provide acceptable means of compliance for the regulatory changes associated with amendments 107, 109, 113, 115, 119, and 123 to part 25, respond to National Transportation Safety Board recommendations, and to provide a general update to reflect current FAA and industry practices and policies.

I.1 Controllability and Maneuverability

Section 3, titled "Controllability and Maneuverability" of AC 25-7C provides the following information and guidance for compliance with § 25.143:

The purpose of § 25.143 is to verify that any operational maneuvers conducted within the operational envelope can be accomplished smoothly with average piloting skill and without encountering a stall warning or other characteristics that might interfere with normal maneuvering, or without exceeding any aircraft structural limits. Control forces should not be so high that the pilot cannot safely maneuver the aircraft. Also, the forces should not be so light that it would take exceptional skill to maneuver the aircraft without over-stressing it or losing control. The aircraft response to any control input should be predictable to the pilot.
The maximum forces given in the table in § 25.143(d) for pitch and roll control for short term application are applicable to maneuvers in which the control force is only needed for a short period. Where the maneuver is

such that the pilot will need to use one hand to operate other controls (such as during the landing flare or a go-around, or during changes of configuration or power/thrust resulting in a change of control force that needs to be trimmed out) the single-handed maximum control forces will be applicable. In other cases (such as takeoff rotation, or maneuvering during en route flight), the two-handed maximum forces will apply.

Short-term and long-term forces should be interpreted as follows:

• Short-term forces are the initial stabilized control forces that result from maintaining the intended flight path following configuration changes and normal transitions from one flight condition to another, or from regaining control following a failure. It is assumed that the pilot will take immediate action to reduce or eliminate such forces by re-trimming or changing configuration or flight conditions, and consequently short-term forces are not considered to exist for any significant duration. They do not include transient force peaks that may occur during the configuration change, change of flight conditions, or recovery of control following a failure.

• Long-term forces are those control forces that result from normal or failure conditions that cannot readily be trimmed out or eliminated.

Compliance with § 25.143 (a) through (g) is primarily a qualitative determination by the pilot during the course of the flight test program. The control forces required and aircraft response should be evaluated during changes from one flight condition to another and during maneuvering flight. The forces required should be appropriate to the flight condition being evaluated. For example, during an approach for landing, the forces should be light and the aircraft responsive in order that adjustments in the flight path can be accomplished with a minimum of workload. In cruise flight, forces and aircraft response should be such that inadvertent control input does not result in exceeding limits or in undesirable maneuvers. Longitudinal control forces should be evaluated during accelerated flight to ensure a positive stick force with increasing normal acceleration. Forces should be heavy enough at the limit load factor to prevent inadvertent excursions beyond the design limit. Sudden engine failures should be investigated during any flight condition or in any configuration considered critical, if not covered by another section of part 25. Control forces considered excessive should be measured to verify compliance with the maximum control force limits specified in § 25.143(d). Allowance should be made for delays in the initiation of recovery action appropriate to the situation.

I.2 Design and Function of Artificial Stall Warning and Identification Systems:

Chapter 8, titled "Design and Function of Artificial Stall Warning and Identification Systems" of AC 25-7C provides the following information and guidance for compliance with Sections 25.103, 25.201, 25.203, and 25.207.

The explanation section of this chapter indicates that some aircraft require artificial stall warning systems to compensate for a lack of clearly identifiable natural aerodynamic stall warning to show compliance with the stall warning requirements of § 25.207. A stick shaker is a recommended method of providing such a warning, regardless of whether or not the natural aerodynamic stall warning is clearly identifiable. Similarly, some aircraft require a stall identification device or system (e.g., stick pusher,) to compensate for an inability to meet the stalling definitions of § 25.201 or the stall characteristics requirements of § 25.203. In addition to compliance with the flight test requirements prescribed in paragraph 29 of this AC, certain system design and function criteria should also be addressed during the certification process of these aircraft. Included are system arming and disarming, preflight checks, failure indications and warnings, and system reliability and safety. The reliability of these systems can be evaluated in terms of the probability of the system not operating when required, and the safety aspects in terms of the probability of the system operating inadvertently. The required reliability and safety of stall warning and identification systems should be defined as a function of how critical their respective functioning is to safety of flight.

The "System Reliability and Safety" section of this chapter indicates the following:

When stall warning and/or stall identification systems are installed to show compliance with the stalling requirements of §§ 25.201, 25.203, and 25.207, engineering data should be supplied to satisfy the following criteria, determined in accordance with § 25.1309.
(1) Reliability. Probability of artificial stall warning and stall identification systems not operating when required:
(a) If stall warning is not clearly identifiable by natural characteristics, the loss of artificial stall warning should be improbable (not greater than 1E-5 per flight hour). This reliability requirement is normally met by using dual, independent stall warning systems.
(b) If the natural stall characteristics are unacceptable, the combination of failure of the stall identification system to operate and entry into a stall should be extremely improbable (not greater than 1E-9 per flight hour). A

stall identification system with a failure rate not greater than 1E-4 per flight hour will satisfy this requirement.

(c) If the stall identification system is installed solely for the purposes of identifying the stall, and the stall characteristics would otherwise meet the requirements of Subpart B with the stall identification system disabled, a maximum failure rate of 1E-3 per flight hour will be acceptable.

(2) Safety. Probability of artificial stall warning and stall identification systems operating inadvertently.

(a) The probability of inadvertent operation of artificial stall warning systems, during critical phases of flight, should not be greater than 1E-5 per flight hour.

(b) To ensure that inadvertent operation of the stall identification system does not jeopardize safe flight, and to maintain crew confidence in the system, it should be shown that:

1 No single failure will result in inadvertent operation of the stall identification system; and

2 The probability of inadvertent operation from all causes is improbable (not greater than 1E-5 per flight hour).

(f) System Functional Requirements.

(1) Operation of the stall identification system should reduce the aircraft angle-of-attack far enough below the point for its activation that inadvertent return to the stall angle-of-attack is unlikely.

(2) The characteristics of stall identification systems, which by design are intended to apply an abrupt nose-down control input (e.g., a stick pusher), should make it unlikely that a flight crew member will prevent or delay its operation. The required stick force, rate of application, and stick travel will depend on the aircraft stall and stick force characteristics, but a force of 50 to 80 pounds applied virtually instantaneously has previously been accepted as providing this characteristic.

(3) Normal operation of the stall identification system should not result in the total normal acceleration of the aircraft becoming negative.

(4) The longitudinal maneuvering capability of an aircraft equipped with stall identification systems, at all speeds likely to be encountered in normal operations, should be substantially the same as would be expected for an aircraft with acceptable aerodynamic stall characteristics.

I.3 AC 25-7 History

On September 26, 1974, FAA Order 8110.8, titled "Engineering Flight Test Guide for Transport Category Airplanes", was published for FAA internal use to describe acceptable means of compliance with the flight test portions of Part 25 of the Federal Aviation Regulations.

On April 9, 1986, the FAA published advisory circular (AC) 25-7, titled, "Flight Test Guide for Certification of Transport Category Airplanes." This new AC indicates that it is an update to FAA Order 8110.8 in the areas of performance and flying qualities covered by subpart B—Flight[52] and the material included in this AC would be removed from Order 8110.8. This new Advisory Circular provided guidelines for the flight test evaluation of transport category airplanes. According to the AC, these guidelines provide an acceptable means of demonstrating compliance with the applicable airworthiness requirements and these methods and procedures have evolved through many years of flight testing of transport category airplanes and, as such, represent current certification practices. Like all AC material, these guidelines are not mandatory and do not constitute regulations. They are derived from previous FAA experience in finding compliance with the airworthiness requirements and represent the methods and procedures found to be acceptable by that experience.

On April 22, 1994, the FAA published a Notice of Proposed Rulemaking (NPRM) 94-15 in the Federal Register (59 FR 19296). In this notice, the FAA proposed amendments to 14 CFR parts 1 and 25 to harmonize certain airworthiness standards for transport category airplanes with the European Joint Aviation Requirements 25 (JAR-25). NPRM 94-15 was developed in response to a petition for rulemaking from the Aerospace Industries Association of America, Inc. (AIA) and the Association Europeenne des Constructeurs de Materiel Aerospatial (AECMA). In their petition, AIA and AECMA requested changes to Section 25.143(c), 25.143(f), 25.149, and 25.201 to standardize certain requirements, concepts, and procedures for certification flight testing and to enhance reciprocity between the FAA and JAA. In addition, the AIA and AECMA recommended changes to FAA Advisory Circular (AC) 25-7, "Flight Test Guide for Certification of Transport Category Airplanes," to ensure that the harmonized standards would be interpreted and applied consistently. The proposals published in NPRM 94-15 were developed by the Aviation Rulemaking Advisory Committee (ARAC) and forwarded to the FAA as an ARAC recommendation. The FAA accepted the recommendation and published NPRM 94-15 for public comment in accordance with the normal rulemaking process.

On June 9, 1995, the FAA published a final rule (72 *Federal Register* Vol. 60, No. 111, Pg. 30743) titled "Revision of Certain Flight Airworthiness Standards to Harmonize with European Airworthiness Standards for Transport Category Airplanes." According to the rule, the FAA is amending part 25 of the Federal Aviation Regulations (FAR) to harmonize certain flight requirements with the European Joint Aviation Requirements 25 (JAR-25). This action responds to a petition from the

Aerospace Industries Association of America, Inc. and the Association Europeenne des Constructeurs de Materiel Aerospatial. These changes are intended to benefit the public interest by standardizing certain requirements, concepts, and procedures contained in the airworthiness standards for transport category airplanes. The effective date of the rule is July 10, 1995.

On March 31, 1998, the FAA released AC 25-7A to update the original AC by incorporating the latest policy and guidance material applicable to all sections of part 25. The material related to regulations outside of subpart B supersedes that contained in Order 8110.8, which has been cancelled accordingly upon issuance of this AC (25-7A). Since AC 25-7 was released on April 9, 1986, it has been the primary source of guidance for flight test methods and procedures to show compliance with the regulations contained in subpart B of part 25, which are related to aircraft performance and handling characteristics. For certification flight testing to show compliance with other part 25 regulations, Order 8110.8, "Engineering Flight Test Guide for Transport Category Airplanes," provided guidance for internal FAA use in determining acceptable means of compliance. Order 8110.8, as revised on September 26, 1974, has been subject to five "change" updates to reflect significant policy changes; the last change being the removal of the subpart B-related material concurrent with the original release of AC 25-7. Order 8110.8 reflected the policy in place when Amendment 25-29 to part 25 was adopted, and the original release of AC 25-7 reflected an Amendment 25-59 time frame. Part 25 has been amended significantly since the two referenced documents were last revised and, likewise, guidance and policy have changed in many areas as experience has been gained.

On March 29, 2011, the FAA released AC 25-7B to add an acceptable means of compliance for the regulatory changes associated with amendments 108, 109, and 115 to part 25, and a revised means of compliance for expansion of takeoff and landing data for higher airport elevations. Means of compliance associated with flight in icing conditions was removed as this material is now contained in AC 25-25.

On October 16, 2012, the FAA released AC 25-7C, which is a significant revision to reduce the number of differences from the European Aviation Safety Agency's Flight Test Guide, provide acceptable means of compliance for the regulatory changes associated with amendments 107, 109, 113, 115, 119, and 123 to part 25, respond to National Transportation Safety Board recommendations, and to provide a general update to reflect current FAA and industry practices and policies.

On May 4, 2018, after certification of the 737 MAX was completed, the FAA released AC 25-7D, to clarify paragraph 23.2.4, Engine Restart Capability—§ 25.903(e); adds paragraph 34.4, Circuit Protective Devices—§ 25.1357; and revises appendix B, Function and Reliability (F&R) Tests, of this AC. This AC has been re-formatted to use a new paragraph numbering system for improved usability. This AC cancels AC 25-7C, *Flight Test Guide for Certification of Transport Category Airplanes*, dated October 16, 2012.

J OVERSIGHT AND DELEGATION

J.1 Inspector General Audit Report

According to a 2011 Office of Inspector General audit report[53], "the FAA is responsible for overseeing numerous aviation activities designed to ensure the safety of the flying public. Recognizing that it is not possible for FAA employees to personally oversee every facet of aviation, public law allows FAA to delegate certain functions, such as approving new aircraft designs, to private individuals or organizations (approved by the FAA). Designees perform a substantial amount of critical work on FAA's behalf—for example, at one aircraft manufacturer, they made about 90 percent of the regulatory compliance determinations for a new aircraft design. FAA created the Organization Designation Authorization (ODA) program in 2005 to standardize its oversight of organizational designees."

According to FAA Order 8100.15A, 49 CFR 44702(d) allows the FAA to delegate to a qualified private person a matter related to issuing certificates, or related to the examination, testing, and inspection necessary to issue a certificate on behalf of the FAA Administrator as authorized by statute to issue under 49 CFR 44702(a).

J.2 Guidance for Delegation of Compliance Findings

FAA Order 8110.4C, section 2.5, titled "Compliance Planning," discusses the FAA's involvement in a certification project, including providing guidance on oversight and delegation. According to the order, "For planning purposes, the FAA's and the applicant's certification teams need to know in which aspects of the project the FAA intends involvement and at what level. The heavy workloads for FAA personnel limit involvement in certification activities to a small fraction of the whole. FAA type certification team members must review the applicant's design descriptions and project plans, determine where their attention will derive the most benefit, and coordinate their intentions with the applicant."

Paragraph (a)(1) of section 2.5 provides guidance to the FAA and applicant on the identification of critical safety items requiring direct FAA involvement in the findings of compliance. According to the paragraph,

"When a particular decision or event is critical to the safety of the product or to the determination of compliance, the FAA must be directly involved (as opposed to indirect FAA involvement by, for example, DER). Project team members must build on their experience to identify critical issues. Some key issues that will always require direct FAA involvement include rulemaking (such as for special conditions), development of issue papers, and compliance findings considered unusual or typically reserved for the FAA. While these items establish the minimum direct FAA involvement, additional critical safety findings must also be identified based on the safety impact or the complexity of the requirement or the method of compliance. Additional factors to consider in determining the areas of direct FAA involvement include the FAA's confidence in the applicant, the applicant's experience, the applicant's internal processes, and confidence in the designees."

6.3 Removed Angle of Attack sensor (P/N 0861FL1;S/N 21401)

Lion Air removed AOA sensor Part Number (P/N) 0861FL1 Serial Number (S/N) 21401 from PK-LQP aircraft on 28 October 2018 to address a maintenance write-up stating that the speed (SPD) and altitude (ALT) flags appeared on the Captain's PFD. Following the accident, BATAM Aero Technik provided the removed AOA sensor to KNKT on 5 November 2018. The KNKT subsequently provided the sensor to the United States National Transportation Safety Board (NTSB) for further examination and detailed testing.

On 10 December 2018, representatives from the KNKT, NTSB, FAA, and Boeing, convened at a Collins Aerospace (previously known as Rosemount Aerospace) facility to perform examination and testing of the AOA sensor in accordance with the Collins Aerospace Component Maintenance Manual (CMM) 34-12-34, Revision 9. Examination of the AOA sensor revealed an intermittent open circuit in the resolver #2 coil wiring. At temperatures above approximately 60°C, the resolver functioned normally, but did not function below that temperature.

The test conducted according to the Rosemount Aerospace Inc., Component Maintenance Manual (CMM) 0861DR, 0861FL, 0816DR MOD1 and 0861FL1 Chapter 34-12-34 Table 1005 Test Data Sheet.

The vane travel test was performed per the test procedure contained within CMM 34-12-34. The AOA sensor passed the requirements for the vane travel at the clockwise (CW) and counter clockwise (CCW) end stops of the unit when the output from Resolver 1 was utilized. The Resolver 1 output is what is used as the requirement within the CMM.

The vane travel was also evaluated using Resolver 2 of this unit. This test would not normally be required within the CMM. The test unit, 0861FL1, SN 21401, failed the requirements for the vane travel at the CW and CCW end stops of the unit. The expected outputs from Resolver 2 were not sinusoidal in nature and therefore the recording instrument could not interpret them. Therefore, the outputs from Resolver 2 were identified on the test data sheet as "unstable."

The resolver accuracy test was performed per the test procedure contained within CMM 34-12-34. The AOA sensor passed the requirements for the resolver accuracy test when the output from Resolver 1 was utilized. The AOA sensor failed the requirements for the resolver accuracy test when the output from Resolver 2 was utilized. The output from Resolver 2 was identified on the test data sheet as "unstable."

The AOA sensor passed the requirements for the heater current test for both the vane and case heaters within the unit.

A resolver accuracy test was repeated on the AOA sensor per the test procedure with the additional requirement to operate the internal heaters. The heater operation was conducted for an appropriate time to ensure proper thermal transfer to the internal components within the unit. This additional test was added outside the normal CMM 34-12-34 requirements to determine if unintended electrical coupling between the heaters and the resolvers was observed.

The first two measurements taken on Resolver 2 showed that the values were unstable similar to values observed in previous resolver accuracy testing. Once the unit warmed up with the heater operation the unit resolver 2 output stabilized and was within the CMM performance requirements. The remaining Resolver 2 values were found within limits. The first two measurements were re-taken and were found within limits. The vane and case heaters were turned off and the values for Resolver 2 went unstable after 12 minutes and 51 seconds. The sine and cosine signals (observed on an oscilloscope) were also being observed and went to zero when the API output went unstable.

The resolver 2 of the AOA sensor part number 08-NCW-24YQ was removed for further examination at the Moog facility in Blacksburg, VA. A physical examination and analysis conducted on June 5-6, 2019 by a group consisting of representatives from NTSB, Boeing, FAA, Collins, and Moog. The failure mode was a temperature dependent intermittent open rotor that eventually failed to operate at lower temperatures.

Using CT Scans, physical examinations, and Scanning Electron Microscope (SEM) imaging, the open circuit was found to be a broken magnet wire on the rotor coil. A loose loop in the coil of the magnet wire had been epoxied between two different insulators on the rotor with different coefficients of thermal expansion for each insulator. As the rotor was exposed to cyclic differences in operating temperature over time, it is likely that the difference in the expansion rates of the two insulators induced a localized stress in the coil wire that led to a fatigue break in the wire that was open or closed dependent on temperature.

As determined by the examination, a loose loop of magnet wire from the primary rotor coil had been trapped with epoxy between the end cap insulator and the rotor shaft insulator. This epoxy is only meant to hold the end cap insulator in place and is not intended to encapsulate the magnet wire. The two insulators are of different material and have different coefficient of thermal expansion (CTE). The end cap insulator has an in plane CTE of 16-20 ppm/°C and the shaft insulator has a CTE of 65 ppm/°C. In addition to the CTE difference between the two insulators, the plane of the shaft insulator is perpendicular to the plane of the end cap insulator. Also note that the rotor coil design necessarily uses very fine magnet wire.

Because of the trapped magnet wire attached to two different CTE materials, the thermal cycling of the resolver over time due to the operational environment of the AOA on the aircraft mechanically "worked" the confined magnet wire into a fatigue failure mode. The magnet wire exhibits a series of ridges or "beach marks" that are indicative of multiple crack growth cycles (i.e. fatigue) before ultimately breaking and arcing multiple times as evidenced in the SEM images.

The examination concluded that the field failure of the 08-NCW-24YQ resolver was due to a loose loop in the rotor coil magnet wire that had been exposed and encapsulated in the epoxy used to hold the end cap insulator on the rotor. The epoxy caused the magnet wire to adhere to both the end cap insulator and the rotor shaft insulator. Because the CTE of the two insulators differ over 3 times from each other, thermal cycling from normal operation in the field caused the magnet wire to fail in fatigue as expansion and contraction rates and possibly directions differed from each side of the magnet wire. The failure manifested as a temperature dependent intermittent open. Physical examination of the resolver, including continuity tests, CT scans, and SEM imaging, concluded that this was the only magnet wire break in the unit and visual evidence of cracking, arcing, and metal "working" support the CTE theory of fatigue of the magnet wire.

APPENDIX 2: Lion Air Comments

COMMENTS ON KNKT FINAL REPORT OF AIRCRAFT ACCIDENT SUBMITTED BY PT LION MENTARI AIRLINES

I. INTRODUCTION

PT Lion Mentari Airlines ("Lion Air") submits the following comments to the KNKT's Final Report on the 29 October 2018 accident involving the Boeing 737-8 MAX aircraft (PK-LQP) operating as Lion Air Flight JT610, which crashed in the Java Sea shortly after takeoff from Soekarno-Hatta International Airport in Jakarta, Indonesia (the "Flight JT610 Accident").

For the reasons discussed below and as explained in: n(1) Lion Air's 18 September 2019 comments to the KNKT's draft Final Report[1] (*see* Exhibit A); (2) the Boeing 737MAX Flight Control System Joint Authorities Technical Review's ("JATR") Observations, Findings and Recommendations dated 11 October 2019 (*see* Exhibit B); 3 the US National Transportation Safety Board's ("NTSB" Safety Recommendation Report dated 19 September 2019 (*see* Exhibit C)) and (4) the public statements of former U.S. Airways Captain Chester "Sully" Sullenberger (*see* Exhibit D),[2] the primary cause of the Flight JT610 Accident was Boeing's defective design and improper self-certification of the 737-8 MAX aircraft-in particular, Boeing;s development and incorporation of the Maneuvering Characteristics Augmentation System ("MCAS"). These deficiencies, combined with Boeing's failure to disclose the existence of MCAS to Lion Air based on its flawed Safety System Analysis ("SSA"), made it impossible for Lion Air's engineers, maintenance personnel, and pilots, including the Flight JT610 pilots, to detect, diagnose, and correct the repetitive, overpowering Aircraft Nose Down (AND) commands by MCAS.

The fatal flaws and flight safety risks that Boeing designed into the 737-8 MAX aircraft also caused the Ethiopian Airlines Flight ET 302 accident on 10 March 2019, less than five months after the Flight JT610 Accident.

H. COMMENTS TO THE FINAL REPORT

  A. Boeing's design and Self Certification of the 737-8 MAX Aircraft, Including MCAS, Caused the Flight JT610 Accident.

In its draft Final Report, KNKT identified numerous design flaws in the Boeing 737-* MAX aircraft, Boeing's failure to implement and conduct the necessary operational safeguards and safety analysis's during its design and developments of MCAS, and the inadequate approval process that enabled Boeing to self-certify MCAS without appropriate oversight. The KNKT also found the Boeing failed to correct a known software error that inhibited the AOA DISAGREE alert from displaying on the pilots' PFD, which would have alerted Lion Air that the left AOA sensor, which was feeding the incorrect date that activated the MCAS AND commands, was malfunctioning

The findings made by the KNKT in its draft Final Report are consistent with the following list of contributing factors that Lion Air has identified as causing the Flight JT610 Accident:

Contributing Factors

- Boeing's design and self-certification of MCAS did not comply with applicable safety and airworthiness standards and permitted a single-point failure-the misaligned left AOA sensor had been mis-calibrated by Xtra Aerospace-to trigger multiple, automatic AND stabilizer trim commands.

- Boeing's MCAS design permitted an unlimited number of AND commands, which resulted in full AND trim that exceeded the elevator authority of the aircraft and overpowered the Flight JT610 crew, and forced the aircraft into an unrecoverable dive.

- Boeing's failure to disclose any information to the pilots about the existence and operation of MCAS during 737-8 MAX differences training and in 737 MAX Operating Manuales and Quick Reference Handbook (QHR) prevented the FlightJT610 crew from properly diagnosing the flight control issues caused by multiple, automatic MCAS AND commands.

- The multiple, contradictory warnings, alerts, and flight control difficulties caused by the single-point failure activation of MCAS was not consistent with a traditional runaway stabilizer event or any other emergency even addressed in the Operating Manuals and QRH and left the Flight JT610 crew without any guidance for correcting the multiple, automatic AND commands. Although unknown by the Flight JT610 crew, the MCAS AND commands could not be countered by control column nose-up movement, normally the first step in the Runaway STAB TRIM memory items, because the control column cutout switches are by-passed by MCAS.

- Boeing's failure to correct a known software error that inhibited the AOA DISAGREE alert from displaying on the PFD during Flight JT43 prevented the Flight JT43 and maintenance personnel at Jakarta from reporting and diagnosing the misalignment of the left AOA sensor before the accident flight.

- Boeing's system safety analyses for MCAS during the Organization Designation Authorization certification process for the 737 MAX failed to account for the deficiencies in MCAS and resulted in Boeing self-certifying an aircraft that was susceptible to causing a "catastrophic event," defined as a failure condition that prevents continued safe flight and landing.

The above list of contributing factors also is supported by the Observations, Finding, and Recommendations issued by the JATR[3] on 11 October 2019, following its review of the type certification of MCAS. See Exhibit B at pp. I-II. Similar to the KNKT, the JATR team found that:

JATR's Findings Related to MCAS

- During its development of MCAS, Boeing did not update certification deliverables "to reflect the changes to this function within the flight control system. In addition (Boeing's) design assumptions were not adequately reviewed, updated or validated; possible flight deck effects were not evaluated; the SSA and functional hazard assessment (FHA) were not consistently updated; and potential crew workload effects resulting from MCAS design changes were not identified." Exhibit B at p. VI.

- (T)the FAA had inadequately awareness of the MCAS function which, coupled with limited involvement, resulted in an inability of the FAA to provide an independent assessment of the adequacy of the Boeing proposed certification activities associated with MCAS. In addition, signs were reported of undue pressured on Boeing ODA engineering unit members (E-Ums) performing certification activities on the B737 MAX program, which further erodes the level of assurance in this system of delegation." Exhibit B at p. VII.

- The JATR team found that the MCAS was not evaluated as a complete and integrated function in the certification documents that were submitted to the FAA. The lack of a unified top-down development and evaluation of the system function and its safety analysis, combined with the extensive and fragmented documentation, made it difficult to assess whether compliance was fully demonstrated. The MCAS design was based on data, architecture, and assumptions that were reused from a previous aircraft configuration without sufficient detailed aircraft-level evaluation of the appropriateness of such reuse, and without additional safety margins and features to address conditions, omissions, or errors not foreseen in the analysis. Exhibit B at p. VIII.

- "During the certification process, a decision was made to remove information relating to MCAS functionality from the draft Flight crew Operating Manual (FCOM). This decision meant that the FAA Flight Standardization Board (FSB) was not fully aware of the MCAS function and was not in a position to adequately assess training needs." Exhibit B at p. XI.

- "The Boeing AFM does not include all the normal, non-normal, and emergency operating procedures as required by regulations." Boeing was able to "make changes to aircraft operating procedures (via the FCOM) without requiring FAA approval for such changes." Exhibit B at pp. XI-XII.

- "If the FAA technical staff had been fully aware of the details of the MCAS function, the JATR team believes the agency likely would have required an issue paper for using the stabilizer in a way that it had not previously been used. MCAS used the stabilizer to change the column force feel, not trim the aircraft. This is a case of using the control surface in a new was that the regulations never accounted for and should have required an issue paper for further analysis by the FAA. If an issue paper had been required, the JATR team believes it likely would have identified for the stabilizer to overpower the elevator" (Exhibit B at pp. 13-14), which was the final failure that lead to the PK-LQP to crash into the Java Sea.

- The JATR team considers that system features on the B737 MAX might constitute a stall identification system. This system is vulnerable to inadvertent actuation due to a single failure, which would not meet the accepted guidance contained within AC 25-7C, Chapter 8, Section 228. Exhibit B at p. 19.

In sum, the JATR found that the Boeing 737 MAX aircraft should not have been certified as airworthy because it was susceptible to uncontrollable AND commands by MCAS based on a single-point failure. This is precisely the catastrophic scenario that caused the Flight JT610 Accident and leads to the conclusion that the Flight JT610 Accident would not have occurred if Boeing;s MCAS system was designed and developed with the appropriate safety protections and feartures, properly certified and approved, and its functionality had been fully explained in the relevant training and operations manuals provided to Lion Air pilots, engineers, and maintenance personnel.

B. Lion Air's Pilots, Engineer, and Maintenance Personnel Should Not Be Faulted For Their Reasonable Attempts to Detect, Diagnose, and Corrects an Unknowable Defect

Lion Air is aware of efforts that have been made to criticize the Lion Air pilots, engineers, and maintenance personnel who operated or worked on the PK-LQP aircraft. Such criticisms are misplaced and should not be considered as contributing factors of the Flight JT610 Accident, as they fail to take into account the circumstances in which these individuals found themselves, having received no training, guidance, or the most basic information about MCAS from Boeing. As the JATR noted, Boeing removed all references to MCAS from its operational manuals, thus making it impossible for Lion Air personnel to detect, diagnose, and correct this latent and unknowable aircraft defect.

1. The Lion Air Flight JT610 crew should not be faulted for their response to a latent and unknowable defect.

With the Boeing 737 MAX aircraft's integrated cockpit and data sharing used by many devices, including MCAS, a single fault or failure involving MCAS cab – without any pilot input or knowledge – rapidly cascade through multiple systems, causing multiple cockpit alarms, cautions and warnings. Such a scenario can lead to distraction, stress, and an exceptional increase in workloads, creating a situation that can quickly become confusing, overwhelming, and ultimately uncontrollable. This is precisely the situation that the Flight JT610 pilots were confronted with shortly after taking off from Jakarta.

Within minutes after takeoff, the failure of a single AOA sensor on Flight JT610 caused multiple instrument indication anomalies and cockpit warnings that signaled numerous potential and conflicting flight control problems. The "speed too slow" warning activated the "stick-shaker," which began and continued to emit rapid and loud vibrations that shook the captain's control column. The "speed too low" warning activated a "clacker," another loud repetitive warning signaling overspeed. Combined, these sudden loud, contradictory, and false warnings created major distractions, interfering with and causing both pilots' focus to shift to the forward panels and the PFD during a critical phase of flight, and making it extremely difficult for any pilot to quickly analyse the situation and take effective corrective action. The compound nature of conflicting warnings and alerts and repeated cycles of AND stab trim was not something the flight crew would have experienced in any prior training or simulator events,[4] as Boeing withheld the existence of and procedures relating to MCAS from the flight crew.

This conclusion is shared by former U.S.Airways Captain Sully Sullengerger, who has flown a 737 MAX full motion simulator that was programmed to replicate both fatal 737 MAX accidents. In a recently disclosed letter, Captain Sullenberger stated:

> **(P)ilots must be capable of absolute mastery of the aircraft and the situation at all times, a concept pilots call airmanship. Inadequate pilot training and insufficient pilot experience are problems worldwide, but they do not excuse the fatally flawed design of the Maneuvering Characteristics Augmentation System (MCAS) that was a death trap. As one of the few pilots who have lived to tell about being in the left seat of an airliner when things went horribly wrong. With seconds to react, I know a thing or two to about overcoming an unimagined crisis. I am also one of the few who have flown a Boeing 737 MAX Level D full motion simulator, replicating both accident flights multiple times. I know firsthand the challenges the pilots on the doomed flights faced, and how wrong it is to blame them for not being able to compensate for such a pernicious and deadly design. The emergencies did not present as a classic runaway stabilizer problem, but initially as ambiguous unreliable airspeed and altitude situations, masking MCAS. The MCAS design should never have been approved, not by Boeing, and not by the Federal Aviation Administration (FAA).**

> *See* Exhibit D (emphasis added).

The NTSB also found that Boeing made incorrect assumptions about the expected pilot response to an unintended MCAS activation. Relying on these assumptions, Boeing failed to asses certain safety scenarios, resulting in an incomplete and flawed SSA.

Specifically, in its Safety Recommendation Report (*see* Exhibit D),[5] the NTSB found that:)

NTSB Safety Recommendation Report

"Multiple alerts and indications can increase pilot's workloas, and the combination of the alerts and indications did not trigger the accident pilots to immediately perform the runaway stabilizer procedure during the initial automatic AND stabilizer trim input. In all three flights (Lion Air Flight JT43, Lion Air Flight JT610, and Ethiopian Airlines Flight ET302), the pilot responses differed and did not match the assumptions of pilot responses to unintended MCAS operation on which Boeing based its hazard classifications within the safety assessment and that the FAA approved and used to ensure the design safely accommodates failures. Although a number of factors, including system design, training, opration, and the pilots' previous experiences, can effect a human's ability to recognize and take immediate, appropriate corrective actions for failure conditions, industry experts generally recognize that an aircraft system should be designed such that the consequences of any human error are limited." Exhibit C at p. 7.

"While Boeing considered the possibility of uncommanded MCAS operation as part of its functional hazard assessment, it did not evaluate all the potential alerts and indications that could accompany a failure that also resulted in uncommanded MCAS operation. Therefore neither Boeing's system safety assessment nor its simulator tests evaluated how the combined effect of alerts and indications might impact pilots' recognition of which procedure(s) to prioritize in responding to an unintended MCAS operation caused by an erroneous AOA output…**Thus, the NTSB concludes that the assumptions that Boeing used in its functional hazard assessment of uncommanded MCAS function for the 737 MAX did not adequately consider and account for the impact that multiple flight deck alets and indications could have on pilots' responses to the hazard.**" Exhibit C at p. 8 (emphasis added).

> "Human factors research has identified that, for non-normal conditions, such as those involving a system failure with multiple alerts, where there may be multiple flight crew actions required, providing pilots with understanding as to which actions must take place priority is a critical need. This is particularly true in the case of functions implemented across multiple airplane systems because a failure in one system within highly integrated system architectures can present multiple alerts and indications to the flight crew as each interfacing system registers the failure. For example, the erroneous AOA output experienced during the two accident flights resulted in multiple alerts and indications to the flight crews, **yet the crews lacked tools to identify the most effective response.**" Exhibit C at pp. 10-11 (emphasis added).

The NTSB correctly emphasizes the impossible situation that the Flight JT610 flight crew faced-not only worth respect to the high workload, stressful, and confusing environment created by the multiple, contradictory warnings in the cockpit, but also that they had no way of knowing that the unintended, repetitive AND commands were coming from MCAS.

2. The flight JT43 crew's post-flight report to maintenance personnel should not be a contributing factor

The KNKT has evaluated the performance of the flight JT43 crew, including the captain's post flight reporting in Jakarta on 28 October 2018. The proper evaluation of the Flight JT43 captain's performance must account for the finding made by the KNKT that the AOA DISAGREE Alert did not appear on the captain's PFD due to a software error, known to Boeing but concealed from operators.

The Flight JT43 captain reported all alerts that appeared on his PFD. It would have been impossible for the Flight JT43 captain to know that the cause of the alerts he saw on the PFD-all of which he noted on the AFML-was the unintended activation of the MCAS as the result of the failure of a single AOA sensor. Stick shaker activation and uncommanded stab trim activation were considered consequences of the speed-related issues that prompted the primary warnings seen on the PFD and forward panel.

Attempting to criticize the Flight JT43 crew's actions in isolation and without context is inappropriate. Neither the Flight JT43 crew nor the Flight JT610 crew had any knowledge, guidance, or training on MCAS.

CONCLUSION

The information obtained by the KNKT during its investigation of the Flight JT610 Accident demonstrates that the primary cause of the Flight JT610 Accident was Boeing's flawed design and development of MCAS. This conclusion is further supported by the independent analyses of Boeing;s certification of the 737-8 MAX aircraft conducted by JATR and the NTSB.

Notes:

[1] The KNKT provided Lion Air with a draft Final Report on 18 August 2019 and a revised draft Final Report on 11 September 2019. Lion Air provided comments tot the KNKT's 11 September draft Final Report on 18 September 2019.

[2] Copies of Exhibits A-D are attached to this submission

[3] JATR was established by the U.S. Federal Aviation Administration ("FAA") is response to the two 737 MAX accidents and consists of technical representatives from the FAA, National Aeronautics and Space Administration ("NASA") and civilization authorities from Australia, Brazil, Canada, China, Europe, Indonesia, Japan, Singapore, and the United Arab Emirates.

[4] Lion Air also suggests that it is inappropriate to attempt to draw conclusions from the pilot's training history in assessing their response to the unintended AND commands that were unknowingly coming from MCAS. Because the Flight JT610 crew had no knowledge of MCAS and received no training on the procedures for responding to multiple MCAS-activated AND commands, the pilots' training history is irrelevant.

[5] The NTSB issued its Safety Recommendation Report, entitled "Assumptions Used in the Safety Assessment Process and the Effects of Multiple Alerts and Indications on Pilots Performance," in connection with its participation in the KNKT's investigation of the Lion Air Flight JT610 Accident and the Aircraft Accident Investigation Bureau of Ethiopia's investigation of the Ethiopian Airlines Flight ET302 Accident. The NTSB's involvement in the investigations prompted it to examine the safety assessments that were conducted as part of the design of Boeing's MCAS on the 737 MAX aircraft.

Notes

[1] Soekarno-Hatta International Airport (WIII), Jakarta will be named as Jakarta for the purpose of this report.

[2] Depati Amir Airport (WIPK), Pangkal Pinang will be named as Pangkal Pinang for the purpose of this report.

[3] The 24-hours clock in Universal Time Coordinated (UTC) is used in this report to describe the local time as specific events occurred. The Local Time (LT) is UTC +7 hours.

[4] Primary Flight Display (PFD) is a dynamic color display of all the parameters necessary for flight path control.

[5] Angle of Attack (AOA) is the angle between wing mean aerodynamic chord and direction of relative wind.

[6] Stick shaker is an artificial warning device to alert the flight crew when airspeed is at a minimum operating speed and is close to a wing stall condition (Boeing 737-8 System Description Section of the Aircraft Maintenance Manual).

[7] The detail of ABASA 1C Standard Instrument Departure (SID) is described in subchapter 1.8 Aids to Navigation.

[8] Waypoint ESALA is located on coordinate 5°57'42.00"S 107°19'0.00"E which about 40 Nm from Soekarno-Hatta Airport on bearing 75°.

[9] Press-on-it-is, is a psychological phenomenon that simply the decision to continue to the planned destination or toward the planned goal even when significantly less risky alternatives exist.

[10] The working time is the time period when the person attends their particular working shift.

[11] The duty time for Air Traffic Controller is the time period when the person performs their duty to provide air traffic control service.

[12] ICAO Annex 6, section 6.5.3.1.

[13] Off block is the time when the aircraft starts pushback or when pushback is not required means when the aircraft starts taxi.

[14] NO GO Item is item(s) of equipment(s) or instrument(s) which unserviceable, failure or absence from an aircraft, prohibits take off according to the regulation or operating rule.

[15] Reference Office of Inspector General Audit Report, AV-2011-136, issued on June 29, 2011.

[16] The Federal Aviation Administration (FAA) uses IPs to provide a structured means to address certain issues in the type certification and type validation processes. Type certification includes projects for TCs, amended TCs, type design changes, STCs and amended STCs. For FAA approvals such as 14 CFR 21.8(d), TSOA, and PMA projects, IPs can be used, with discretion, to document and resolve compliance issues where directorate or policy office guidance is required. Reference FAA Advisory Circular (AC) 20.166.

[17] A Type Certificate Data Sheet (TCDS) is a formal description of the aircraft, engine or propeller. It lists limitations and information required for type certification including airspeed limits, weight limits, thrust limitations, etc.

[18] The hazard assessments were developed as part of aircraft certification and based on AC 25.1309-1A.

[19] The two events were assumed to start from a trimmed condition. Boeing also considered the hazard of uncommanded MCAS operation until pilot response. This condition had the same severity as the 3-second case.

[20] FAA advisory circular (AC) 25-7C, titled, "Flight Test Guide for Certification of Transport Category Airplanes," dated October 16, 2012, provides guidance for the flight test evaluation of transport category airplanes.

[21] FAR 25.143(g) Controllability and Maneuverability – General, Requires that changes of gradient that occur with changes of load factor must not cause undue difficulty in maintaining control of the airplane, and local gradients must not be so low as to result in a danger of over-controlling. Reference is made to CFR amendment 25-129 for the described FAR 25.143(g) requirement.

[22] According to the ICAO Document 9859, hazard is a condition or an object with the potential to cause or contribute to an aircraft incident or accident.

[23] I Gusti Ngurah Rai International Airport (WADD), Denpasar will be named as Denpasar for the purpose of this report.

[24] Takeoff Configuration Warning is a safety device intended to ensure that takeoff is properly configured. An intermittent warning horn and the TAKEOFF CONFIG warning light illuminates when takeoff configuration warning activates.

[25] Primary Flight Display (PFD) is primary reference for flight information which displays electromechanical instruments onto a single electronic display.

[26] Reduced Vertical Separation Minima (RVSM) airspace is any airspace or route between altitude 29,000 feet (flight level (FL) 290) and altitude 41,000 (FL 410) inclusive where aircraft are separated vertically by 1,000 feet. The aircraft shall be equipped with two independent altitude measurement systems that meets RVSM performance requirements, an altitude alerting system, and an automatic altitude control system (a certified autopilot that is used while in the RVSM environment) that is sufficient to comply with RVSM performance requirements and several other requirements

[27] Both the 737-8 and 737-9 were in service at the time of the accident. The 737-7 and 737-10 are planned future derivatives that have not yet entered service.

[28] The flight control computers (FCC) are part of the digital flight control system. There are two autopilots, autopilot A from FCC A and autopilot B from FCC B.

[29] When the flaps are down, the stabilizer rate is three times faster than when the flaps are up.

[30] Two stabilizer trim cutout switches on the control stand can be used to stop the main electric and autopilot trim inputs to the stabilizer trim actuator. The switches can be set to NORMAL or CUTOUT. If either switch is moved to CUTOUT, both the electric and autopilot trim inputs are disconnected from the stabilizer trim motor. NORMAL is the default position to enable operation of the electric and autopilot trim.

[31] MCAS failures do allow the stabilizer to move at the flaps down trim rate, even if the flaps are up, but even the flaps down trim rate is a limit, albeit faster than the normal flaps up rate. Column cutout is always available in the forward direction but may not be available in the aft direction for certain MCAS failures.

[32] MCAS is an open loop flight control law.

[33] The safety analysis contained two sections that discussed hazard analysis; the first FHA was developed for the 737NG in the original release of the analysis (1997) and the second FHA was developed as part of the 737 MAX changes (2016).

³⁴ The hazard assessments were developed as part of aircraft certification and based on AC 25.1309-1A.

³⁵ The two events were assumed to start from a trimmed condition. Boeing also considered the hazard of uncommanded MCAS operation until pilot response. This condition had the same severity as the 3-second case.

³⁶ FAA advisory circular (AC) 25-7C, titled, "Flight Test Guide for Certification of Transport Category Airplanes," dated October 16, 2012, provides guidance for the flight test evaluation of transport category airplanes.

³⁷ Boeing's Flight Controls –Autoflight (EDFCS/FCC) & Autothrottle Certification Plan CP13474 indicated that the software will be developed by Rockwell Collins and the Software Accomplishment Summary (SAS) document will be a summary of all the design development and verification activities defined in the PSAC that provides the data to substantiate that the objectives of RTCA DO-178B for the appropriate design software level have been met.

³⁸ FAR 21.20 is the provision setting forth the responsibility for showing compliance. Applicable Orders are 8110.4 (Type Certification) and 8110.15 (ODA)

³⁹ Certification accounts for proper completion of tasks established for flight operations and ground crew maintenance tasks and it relies on decision making and actions being based on an established safety culture.

⁴⁰ Meeting held at the FAA on February 27, 2019.

⁴¹ Meeting held at the FAA on February 27, 2019.

⁴² A Type Certificate Data Sheet (TCDS) is a formal description of the aircraft, engine or propeller. It lists limitations and information required for type certification including airspeed limits, weight limits, thrust limitations, etc.

⁴³ The FAA accepts certification plans; it does not approve the plans.

⁴⁴ The FHA was included in CP13471 beginning at Rev NEW based on the revision history. There were updates made to the FHA in subsequent revisions and the final System Safety Analysis accurately reflects the FHA classifications.

⁴⁵ Deliverables are documents to be submitted demonstrating compliance with the applicable requirements.

⁴⁶ Outside Boeing Authorized Representative – An individual acting under the authority of the Boeing ODA who is not employed by Boeing. The name has since changed "Outside Boeing Engineering Unit Member".

⁴⁷ The basic SSEC factor is a polynomial equation using a combination of airplane Mach, AOA and airframe-specific coefficients ("local measurement to aircraft true correction coefficients") which are established during wind tunnel and flight testing.

⁴⁸According to Boeing, their records indicated that the briefing was originally scheduled for May 2016 and documents were provided to the FAA. However, their records indicated that the actual briefing was delayed and did not take place until July 2016.

⁴⁹ Note the two different flight envelopes designated for MCAS related hazards – "Normal Flight Envelope" and "Operating Flight Envelope". Operating flight envelope is defined in FAA AC 25-7C Appendix 5. There is a difference between flight phase and flight envelope. Phases would be takeoff, climb, cruise, etc. Envelopes are related to altitude, weight, airspeed, AOA, maneuvering "g" loads, etc.

[50] A failure condition of an erroneous MCAS activation preventing the column cutout mechanism from interrupting (as designed) an uncommanded nose-down automatic stabilizer trim command

[51] According to Boeing, because the change in MCAS authority did not change the FHA category of uncommanded MCAS, there was no reason to revisit the S&MF analysis.

[52] Reference item 2 of Advisory Circular 25-7, dated April 9, 1986.

[53] Reference Office of Inspector General Audit Report, AV-2011-136, issued on June 29, 2011.

Other Air Crash Investigations:

Adam Air Flight DHI 574,
On 1 January 2007, a Boeing 737-4Q8, operated by Adam Air as flight DHI 574, was on a flight from Surabaya, East Java to Manado, Sulawesi, at FL 350 (35,000 feet) when it suddenly disappeared from radar. There were 102 people on board. Nine days later wreckage was found floating in the sea near the island of Sulawesi. The black boxes revealed that the pilots were so engrossed in trouble shooting the IRS that they forgot to fly the plane, resulting in the crash that cost the lives of all aboard.

Aeroflot Flight 821
On 14 September 2008 Aeroflot Flight 821, a Boeing 737-505, operated by Aeroflot-Nord, a subsidiary of the Russian airline Aeroflot, crashed on approach to Bolshoye Savino Airport, Perm, Russia. All 82 passengers and 6 crew members were killed. The aircraft was completely destroyed. According to the final investigation report, the main reason of the crash was pilot error. Both pilots had lost spatial orientation, lack of proper training, insufficient knowledge of English and fatigue from lack of adequate rest. Alcohol in the Captain's blood may also have contributed to the accident.

Aerospatiale (Eurocopter) SA365N1, the crash of a helicopter
On September 27, 2008, an Aerospatiale (Eurocopter) SA365N1, call sign Trooper 2, operated by the Maryland State Police (MSP) as a public medical evacuation flight, encountered b ad weather en route to the hospital and was diverted to Andrews Air Force Base (ADW), Camp Springs, Maryland, 3.2 miles north of the runway 19R threshold at ADW, the helicopter impacted terrain and crashed. The commercial pilot, one flight paramedic, one field provider, and one automobile accident patients being transported were killed. The NTSB blamed the pilot's inexperience for the crash.

Air Algérie Flight DAH 6289
On Thursday 6 March 2003, Air Algérie Flight DAH 6289, a Boeing 737-200, suffered during take-off from Tamanrasset, in Southern Algeria, a contained burst in the left engine. The airplane swung to the left, lost speed progressively, stalled and crashed, with the landing gear still extended, about one thousand six hundred and forty-five meters from the takeoff point, to the left of the runway extended centerline. The crew and 96 passengers were killed in the accident, one passenger survived. The airplane was on a domestic flight from Tamanrasset to Ghardaïa and Algiers.

Air Asia Indonesia Flight 8501
On 28 December 2014 an Airbus A320-216 aircraft from Air Asia Indonesia was on a flight from Surabaya, Indonesia to Singapore with total occupants of 162 persons. The Flight Data Recorder (FDR) recorded that many master cautions activated following the failure of the Rudder Travel Limiter. The aircraft stalled and crashed. The investigation showed that the loss of electricity and the RTLU

failure were caused by a cracked solder joint. All occupants of the plane were killed in the accident.

Air China Flight 129
On April 15, 2002, about 11:21:17, Air China flight 129, a Boeing 767-200ER, operated by Air China, en route from Beijing, China to Busan, Korea, crashed on Mt. Dotdae, near Gimhae Airport, Busan. Of the 166 persons on board, 37 persons survived the crash, while 129 occupants were killed. The probable cause of the crash was pilot and ATC errors, while the airport did not inform the captain of the bad weather conditions at the time of landing in Busan. Because of these conditions eight previous flights were diverted to other airports.

Air Florida Flight 90
On January 13, 1982, Air Florida Flight 90, a Boeing 737-222, was a flight to Fort Lauderdale, Florida, from Washington National Airport, Washington, D.C. There were 79 people on board. The flight was delayed due to snowfall. Shortly after takeoff the aircraft crashed into the Potomac River, 74 people in the plane and 4 on the bridge were killed. The National Transportation Safety Board determines that the probable cause of this accident was the flight crew's misinterpretation of the weather conditions. Contributing were the ground delay between de-icing and takeoff clearance.

Air France Flight 358
On August 2, 2005 Air France Flight 358, an Airbus A340, departed Paris, on a flight to Toronto, Canada, with 297 passengers and 12 crew members on board. On final approach, the aircraft's weather radar was displaying heavy precipitation. The aircraft touched down 3800 feet down the runway, and was not able to stop before the end of it. The aircraft stopped in a ravine and caught fire. All passengers and crew members were able to evacuate the aircraft on time. Only 2 crew members and 10 passengers were seriously injured during the crash and the evacuation.

Air France Flight AF447
On 31 May 2009, flight AF447, an Airbus A330-200, took off from Rio de Janeiro bound for Paris. At 2 h 10, a position message and some maintenance messages were transmitted by the ACARS automatic system. After this nothing was heard of from the aircraft. Six days later bodies and airplane parts were found by the French and Brazilian navies. All 228 passengers and crew members on board are presumed to have perished in the accident. A massive search by air and sea craft for the plane's black boxes failed so far.

Air France Flight AF 447 (final report)
On 31 May 2009, the Airbus A330 flight AF 447 took off from Rio de Janeiro Galeão airport bound for Paris Charles de Gaulle. At 2h 10min 05, likely following the obstruction of the Pitot probes by ice crystals, the speed indications became incorrect and some automatic systems disconnected, the aero plane came in a stall situation and crashed in the sea at 2 h 14 min 28s, killing all 228 persons

on board. It took two years to recover the wreck of the aircraft from a depth of 4.000 meters. After two intermediate reports this is the final report of the crash.

Air France Flight AFR 4590
On Tuesday 25 July 2000 Air France Flight AFR 4590, a Concorde registered F-BTSC, took off from Paris Charles de Gaulle, to undertake a charter flight to New York with nine crew members and one hundred passengers on board. During takeoff from runway 26 right at Roissy Charles de Gaulle Airport, a tire was damaged. A major fire broke out. The aircraft was unable to gain height or speed and crashed onto a hotel, killing all 109 people on board and 4 on the ground. The crash would become the end of the Concorde era.

Air India Flight 182
On 23 June 1985, Air India Flight 182, a Boeing 747-237B was on its way from Montreal, Canada, to London when it was blown up while in Irish airspace, and crashed into the Atlantic Ocean. 329 people perished. It was the largest mass murder in modern Canadian history. The explosion and downing of the carrier was related to the Narita Airport Bombing. Investigation and prosecution took 25 years. The suspects were members of the Sikh separatist Babbar Khalsa. Inderjit Singh Reyat, the only person convicted, was sentenced to 15 years in prison.

Air Transat Flight 236
On August 24, 2001, Air Transat Flight 236, an Airbus 330, was on its way from Toronto, Canada to Lisbon, Portugal with 306 people on board. Above the Atlantic Ocean, the crew noticed a dangerous fuel imbalance. After flying 100 miles without fuel the crew managed to land the aircraft at the Lajes Airport at 06:45. Only 16 passengers and 2 cabin-crew members received injuries. The investigation uncovered a large crack in the fuel line of the right engine, caused by mistakes during an engine change just before the start of the flight.

Alaska Airlines Flight 261
On January 31, 2000, Alaska Airlines, Flight 261, a McDonnell Douglas MD-83, was on its way from Puerto Vallarta, Mexico, to Seattle, Washington, when suddenly the horizontal stabilizer of the plane jammed. Captain Thompson and First officer Tansky tried to make an emergency landing in Los Angeles. The plane suddenly crashed into the Pacific Ocean, killing all 93 people aboard. The NTSB concluded that the crash was caused by insufficient maintenance. The crash of Alaska Airlines Flight 261 could have been avoided.

American Airlines Flight 191
On May 25, 1979, American Airlines Flight 191, a McDonnell-Douglas DC-10-10 aircraft, on its way from Chicago to Los Angeles, crashed just after take-off near Chicago-O'Hare International Airport, Illinois. During the take off the left engine and pylon assembly and about 3 ft. of the leading edge of the left wing separated from the aircraft and fell to the runway. Flight 191 crashed killing two hundred and seventy one persons on board and two persons on the ground. The accident remains the deadliest airliner accident to occur on United States soil.

American Airlines flight 587
On November 12, 2001, American Airlines flight 587, an Airbus A300-605R, took off from John F. Kennedy International Airport, New York. Flight 587 was a scheduled passenger flight to Santo Domingo, Dominican Republic, with a crew of 9 and 251 passengers aboard the airplane. Shortly after take-off the airplane lost its tail, the engines subsequently separated in flight and the airplane crashed into a residential area of Belle Harbor, New York. All 260 people aboard the airplane and 5 people on the ground were killed, and the airplane was destroyed by impact forces and a post crash fire.

American Airlines Flight 625
On April 27, 1976, American Airlines Flight 625, a Boeing 727, during landing at the Harry S Truman Airport, Charlotte Amalie, St. Thomas, Virgin Islands, overran the end of runway 9, struck the ILS antenna, crashed through a fence, and hit a building located about 1,040 feet beyond the departure end of the runway. The aircraft was destroyed, 35 passengers and 2 flight attendants were killed. The National Transportation Safety Board determined that the probable cause of the accident was the captain's actions and his misjudgment during landing.

American Airlines Flight 965
On December 20, 1995, American Airlines Flight 965, a Boeing 757-223, was on a scheduled passenger flight from Miami, Florida, U.S.A., to Cali, Colombia. Close to its final destination the pilots erroneously cleared the approach waypoints from their navigation computer. When the controller asked the pilots to check back in over Tuluá, north of Cali, it was no longer programmed into the computer. They were lost and the aircraft crashed into a mountain. Of the 163 people on board, 4 passengers survived the accident.

Armavia Flight RNV 967
On 2 May 2006 Armavia Flight RNV 967, an Airbus A320, was as a passenger flight on its way from Yerevan, Armenia to Sochi, Russia. Upon approaching Sochi there was confusion in regard to the cloud level and visibility for the scheduled landing. While trying to land the air traffic control ordered the captain to abort the landing. In the attempt to abort the captain lost control and the aircraft crashed in the sea, killing all 113 people on board. Contributing factors were not maintaining strict cockpit rules and deficiencies by air traffic control.

Atlantic Southeast Airlines flight 529
This book explains the accident involving Atlantic Southeast Airlines flight 529, an EMB-120RT airplane, which lost a propeller blade and crashed near Carrollton, Georgia, on August 21, 1995. The accident killed 8 people on board. Safety issues in the report focused on manufacturer engineering practices, propeller blade maintenance repair, propeller testing and inspection procedures, the relaying of emergency information by air traffic controllers, crew resource management training, and the design of crash axes carried in aircraft.

British Airways Flight BA38
On 28 November 2008, a Boeing 777-200ER, operated by British Airways as flight BA38, on its way from Beijing, China, to London (Heathrow), suffered on approach to Heathrow Airport an in-flight engine rollback. At 720 feet agl, both engines ceased responding to auto throttle commands. The result was that the aircraft touched down 330 m short of the paved surface of Runway 27L at London Heathrow. The reduction in thrust was due to restricted fuel flow to both engines, caused by the forming of ice in the fuel system. The aircraft was destroyed, but there were no casualties.

China Eastern Airlines Flight 583,
On April 6, 1993, a China Eastern Airlines McDonnell Douglas MD-11, flight 583, on its way from Beijing, China, to Los Angeles, California, had an inadvertent deployment of the leading edge wing slats while in cruise flight, not far from Shemya, Alaska. The captain was manually controlling the airplane when it progressed through several violent pitch oscillations and lost 5,000 feet of altitude. Two passengers were fatally injured, and 149 passengers and 7 crewmembers received various injuries. The probable cause of this accident was the inadequate design of the flap/slat actuation handle.

Colgan Air Flight 3407,
On February 12, 2009, about 2217 Eastern Standard Time, Colgan Air, Flight 3407, a Bombardier DHC-8-400, on approach to Buffalo-Niagara International Airport, crashed into a residence in Clarence Center, New York, 5 nautical miles northeast of the airport. The 2 pilots, 2 flight attendants, and 45 passengers aboard the airplane were killed, one person on the ground was killed, and the airplane was destroyed. The National Transportation Safety Board determined that the probable cause of this accident was a pilot's error.

Comair Flight 5191
On August 27, 2006, Comair Flight 5191, a Bombardier CL-600-2B19, crashed during takeoff from the wrong runway of Blue Grass Airport, Lexington, Kentucky, killing 49 of the 50 people aboard. From the beginning everything went wrong. First the captain and first officer boarded the wrong airplane, only after starting the auxiliary power unit they found out they were in the wrong aircraft. Taxiing to the takeoff position the captain and first officer were so engaged in a private conversation that they did not realize they took the wrong runway. The air traffic controller did not notice anything.

Crash of John F. Kennedy Jr.
At 8:39 p.m. on July 16,1999 John F. Kennedy Jr., son of former President John F. Kennedy and Jacqueline Lee Bouvier, departed from Essex County Airport (IATA: CDW), a public use airport located in Fairfield Township, Essex County. His destination was Martha's Vineyard Airport. There were three persons on board the Piper Saratoga: John F. Kennedy Jr, his wife, and sister-in-law. Kennedy's plane never arrived in Martha's Vineyard. The plane wreck was

discovered in the Atlantic Ocean two days later. All three inhabitants of the plane were dead. The cause of the crash was spatial disorientation.

Crash of Sikorsky S-76A Helicopter G-BJVX
On March 23, 2004, about 1918:34 central standard time, an Era Aviation Sikorsky S-76 helicopter, crashed into the Gulf of Mexico about 70 nautical miles south-southeast of Scholes International Airport (GLS), Galveston, Texas. The helicopter was en route to the drilling ship Discoverer Spirit. All 10 people aboard the helicopter were killed. The helicopter was destroyed. The probable cause of this accident was the flight crew's failure to identify and arrest the helicopter's descent which resulted in controlled flight into terrain.

Delta Air Lines Flight 1288
On July 6, 1996, a McDonnell Douglas MD-88, N927DA, operated by Delta Air Lines Inc., as flight 1288, experienced an engine failure during the initial part of its takeoff roll on runway 17 at Pensacola Regional Airport in Pensacola, Florida. Uncontained engine debris from the front compressor front hub (fan hub) of the No. 1 (left) engine penetrated the left aft fuselage. Two passengers were killed and two others were seriously injured. The takeoff was rejected, and the airplane was stopped on the runway.

Eastern Air Lines Flight 401
On December 29, 1972 an Eastern Air Lines' Lockheed L-1011, as Flight 401 on its way from John F. Kennedy International Airport, New York, to Miami International Airport, Miami, Florida, crashed at 2342 EST in the Everglades, just west-northwest of Miami International Airport. The aircraft was destroyed, 99 people 99 died in the crash. The flight had problems with the nose landing gear. The National Transportation Safety Board determines that the probable cause of this accident, was that preoccupation with a malfunction distracted the crew and allowed the descent to go unnoticed.

EgyptAir Flight 990 (1)
On October 31, 1999, EgyptAir flight 990, a Boeing 767-366ER crashed into the Atlantic Ocean 60 miles south of Nantucket, Massachusetts. All 217 people on board were killed, and the airplane was destroyed. The US National Transportation Safety Board determines that the probable cause of the accident is the airplane's departure from normal cruise flight and subsequent impact with the Atlantic Ocean as a result of the relief first officer's flight control inputs. The reason for the relief first officer's actions was not determined.

EgyptAir Flight 990 (2)
On October 31, 1999, EgyptAir flight 990, a Boeing 767-366ER crashed into the Atlantic Ocean 60 miles south of Nantucket, Massachusetts. All 217 people on board were killed, and the airplane was destroyed. Contrary to the conclusions of the American NTSB the Egyptian Investigation Team concludes that a mechanical defect is the most likely cause of the accident. According to the

Egyptians there is no evidence to support a conclusion that the First Officer intentionally dove the airplane into the ocean in fact.

El Al Israel Airlines Flight 1862
On 4 October 1992, El Al Israel Airlines Flight 1862, a Boeing 747-200 Freighter, departed from Schiphol Airport, Amsterdam, on its way to Tel Aviv, Israel. Seven minutes after take-off the plane lost engine no. 3 and 4 and crashed in an apartment block just outside Amsterdam, killing 43 people (4 crewmembers and 39 on the ground). The investigation concluded that the design and certification of the B 747 pylon was inadequate to provide the required level of safety. Furthermore the system to ensure structural integrity by inspection failed.

Ethiopian Airlines Flight ET 409
On 25 January 2010, at 00:41:30 UTC, Ethiopian Airlines flight ET 409, a Boeing 737-800, on its way from Beirut to Addis Abeba, crashed just after take-off from Rafic Hariri International Airport in Beirut, Lebanon, into the Mediterranean Sea about 5 NM South West of Beirut International Airport. All 90 persons onboard were killed in the accident. The investigation concluded that the probable causes of the accident were pilot errors due to loss of situational awareness. Ethiopian Airlines refutes this conclusion.

Eurocopter AS350BA helicopter collides with Piper PA-32R-300
On August 8, 2009, at 1153:14 eastern daylight time, a Piper PA-32R-300 airplane, and a Eurocopter AS350BA helicopter, collided over the Hudson River near Hoboken, New Jersey. The pilot and two passengers aboard the airplane and the pilot and five passengers aboard the helicopter were killed, and both aircraft were completely destroyed from the impact. According to the National Transportation Safety Board the probable cause of this accident was the inherent limitations of the see-and-avoid concept, and the distraction of the Teterboro Airport local controller.

Federal Express Flight 1478
On July 26, 2002, about 0537 eastern daylight time, Federal Express flight 1478, a Boeing 727-232F, on its way from Memphis International Airport to Tallahassee Regional airport, struck trees on short final approach and crashed short of runway 9 at the Tallahassee Regional Airport, Florida. The captain, first officer, and flight engineer were seriously injured, and the airplane was destroyed. The National Transportation Safety Board determines that the probable cause of the accident was the crew's failure to establish and maintain a proper glide path during the approach to landing.

Germanwings Flight 9525
On Tuesday 24 March 2015, the Airbus A320-211 registered D-AIPX operated by Germanwings took off from Barcelona, Spain, at 09:00 with destination Düsseldorf, Germany. At 09:41, the aircraft crashed into the mountains northeast of Marseille. The investigation into the causes of the crash revealed that the co-

pilot, at a moment when he was alone in the cockpit, had deliberately flown the plane into the mountains killing all 150 persons on board. The investigation revealed also that the co-pilot was under medical treatment for depressions by several health care providers.

Helios Airways Flight 522

On 14 August 2005, a Boeing 737-300 aircraft departed from Larnaca, Cyprus, for Prague. As the aircraft climbed through 16.000 ft., the Captain contacted the company Operations Centre and reported a problem. Thereafter, there was no response to radio calls to the aircraft. At 07:21 h, the aircraft was intercepted by two F-16 aircraft of the Hellenic Air Force. They observed the aircraft and reported no external damage. The aircraft crashed approximately 33 km northwest of the Athens International Airport. All 121 people on board were killed.

Japan Airlines Flight 123

On August 12, 1985, a Japan Airlines B-747 aircraft lost, shortly after take-off, part of its tail and crashed in the mountains northwest of Tokyo. Of the 524 persons on board 520 were killed, 4 survived the accident. The accident was caused by a rupture of the aft pressure bulkhead of the aircraft, and the subsequent ruptures of a part of the fuselage tail, vertical fin and hydraulic flight control systems. The rupture happened as the result of an improper repair after an accident with the aircraft in Osaka, in June 1978.

Kenya Airways Flight KQA 507

During the night of 04th May 2007, the B737-800, registration 5Y-KYA, operated by Kenya Airways as flight KQA 507 from Abidjan international airport (Ivory Coast), to the Jomo Kenyatta airport Nairobi (Kenya), made a scheduled stop-over at the Douala international airport (Cameroon). Shortly after take-off at about 1000 ft., the aircraft entered into a slow right roll that increased continuously and eventually ended up in a spiral dive, the airplane crashed in a mangrove swamp, killing all 114 people aboard. The captain, as pilot flying, lost complete control.

Kish Airlines Flight IRK 7170

On February 10, 2004, Kish Airlines Flight IRK 7170, a Fokker F27, operated a scheduled passenger flight from Kish Island, Islamic Republic of Iran, to Sharjah, United Arab Emirates, a 35 minutes flight. During the approach of Sharjah International Airport, the aircraft was observed to pitch down and suddenly turn to the left. The aircraft continued to descend and turn at high pitch and roll angles and impacted a sandy area 2.6 nm from the runway threshold. A large explosion was seen. The aircraft was destroyed and there were 43 fatalities.

KLM Flight 4805 and PANAM Flight 1736

On Sunday, March 27, 1977 KLM Flight 4805 and PANAM Flight 1736 both approached Las Palmas Airport in the Canary Islands, when a terrorist's bomb exploded on the airport. Both flights were diverted to the neighboring island of Tenerife. After Las Palmas Airport reopened first KLM Flight 4805 was cleared for takeoff, a few minutes later PANAM 1736 was cleared. Due to a number of misunderstandings both aircraft collided on the runway of Tenerife Airport during takeoff, killing 583 people.

Korean Air Flight 801
On August 6, 1997, about 0142:26 Guam local time, Korean Air flight 801, a Boeing 747-300, crashed at Nimitz Hill, Guam. The aircraft was on its way from Seoul, Korea to Guam with 237 passengers and a crew of 17 on board. Of the 254 persons on board, 228 were killed. The airplane was destroyed by impact forces and a post-crash fire. The National Transportation Safety Board determined that the probable cause of the accident was captain's fatigue and Korean Air's inadequate flight crew training.

Lauda Air Flight NG 104
Lauda Air Flight NG 104, a Boeing 767-300 ER of Austrian nationality was on a scheduled passenger flight Hong Kong-Bangkok-Vienna, Austria. NG 104 departed Hong Kong Airport on May 26, 1991, and made an intermediate landing at Bangkok Airport. The flight departed Bangkok Airport at 1602 hours. The airplane disappeared from air traffic radar at 1617 hours, about 94 nautical miles northwest of Bangkok. The probable cause of this accident is attributed to an uncommanded in-flight deployment of the left engine thrust reverser. All 223 people on board died in the accident.

Malaysia Airlines Flight MH370
On 07 March 2014 at 1642 UTC, a Malaysia Airlines Flight MH370, bound for Beijing departed from Kuala Lumpur International Airport with 239 persons on board. It was a Boeing 777-200ER. Suddenly all communication stopped and the plane changed course to the South Indian Ocean. On 1 August 2015 a piece of the wing was found. The accident is very similar to the crash of Helios Flight 5223 on 13 August 2005. This plane suffered from cabin pressure loss. What is the role of the American base in Diego Garcia in the story?

Northwest Airlines Flight 255,
About 2046 eastern daylight time on August 16, 1987, Northwest Airlines flight 255, a McDonnell Douglas DC-9-82, a passenger flight en route to Phoenix, Arizona, crashed shortly after taking off at the Detroit Metropolitan Wayne County Airport, Romulus, Michigan. Of the 155 people on board passengers only a 4-year-old child, survived... On the ground, two persons were killed. The National Transportation Safety Board determines that the probable cause of the accident was the flight-crew's failure to use the taxi checklist.

Olympic Airlines Flight OA202
The Lockheed 1011 registered A6-BSM, chartered by Olympic Airlines, left as flight OA202 Charles de Gaulle International Airport, Paris, on 4 July 2005 at 16h18. Shortly after departure there were engine problems. The captain returned immediately. An investigation by the French BEA showed numerous technical problems with the aircraft, such as fuel and hydraulic leakages, non-working fire

alarms and lack of maintenance. The flight crew was not properly licensed and the insurance was insufficient. In fact A6-BSM was a flying coffin.

PANAM Flight 103
On the 21st of December 1988, PANAM Flight 103, a Boeing 747-121, on its way from London Heathrow to New York, was blown up over the town of Lockerbie, Scotland. All 259 persons on board of the aircraft and 11 residents of the town of Lockerbie were killed. In 2001 the Libyan Megrahi was sentenced to life imprisonment in Scotland. In 2009 Megrahi applied to be released from jail on compassionate grounds. His appeal was granted and on the 20th of August 2009 he was released from prison. But was Megrahi really guilty?

Polish Flight PLF 101
On April 10, 2010 at 10:41 local time, approaching Runway 26 of Smolensk "Severny" airdrome, a Tupolev-154M aircraft of the State Aviation of the Republic of Poland crashed while conducting a non-regular international flight PLF 101 carrying passengers from Warsaw to Smolensk. The cause of the accident was the failure of the crew to take a timely decision to proceed to an alternate airdrome due to weather conditions at the airport of destination. All 96 persons on board, including Polish President Lech Kaczyński and his wife, died in the crash.

Sibir Airlines Flight C7 778
On July 8, 2006 an A-310 airplane operated by Sibir Airlines, ran down the runway, overran the runway threshold and collided with barriers, broke apart and burst into flames. As a result of the accident 125 individuals died, including both pilots and 3 of the cabin crew; 60 passengers and 3 cabin crew suffered injuries. The actions of the crew revealed shortcomings in the professional training. The real cause of the accident was therefore pilot error due to lack of training and experience.

SilkAir Flight 185
On 19 December 1997 SilkAir Flight 185, a Boeing 737-300, operated by SilkAir, Singapore, on its way from Jakarta to Singapore, crashed at about 16:13 local time into the Musi river near Palembang, South Sumatra. All 97 passengers and seven crew members were killed. Prior to the sudden descent from 35,000 feet, the flight data recorders suddenly stopped recording at different times. There were no mayday calls transmitted from the airplane prior or during the rapid descent. The weather at the time of the crash was fine.

Spanair Flight JKK5022
On 20 August 2008, Spanair flight JKK5022, a McDonnell Douglas DC-9-82 (MD-82), crashed during take-off from Barajas Airport in Madrid, The investigation revealed that the accident occurred as the aircraft attempted to take off, because the pilots had omitted to deploy the flaps and slats ready for take-off. The MD-82 warning system, that should have alerted the pilots that the plane was

incorrectly configured for take-off, did not sound a warning. Of the 172 people on board 154 perished in the accident. Most burned alive.

Swissair Flight SR 111

On 2 September 1998, Swissair Flight SR 111 departed New York, flight to Geneva, Switzerland, with 215 passengers and 14 crew members on board. About 53 minutes after departure, the flight crew smelled an abnormal odor in the cockpit. They decided to divert to the Halifax International Airport. They were unaware that a fire was spreading above the ceiling in the front area of the aircraft. They did not make it to Halifax, 20 minutes later the aircraft crashed in the North Atlantic near Peggy's Cove, Nova Scotia, Canada. There were no survivors, 229 people died in the incident.

TAM Linhas Aéreas Flight JJ3054

On 17 July 2007 an Airbus A-320, operated as flight JJ3054 by the Brazilian company TAM Linhas Aéreas, was on its way from Porto Alegre, Brazil, for a domestic flight to Congonhas Airport in São Paulo city, Brazil. During the landing, the aircraft was not slowing down as expected, veered to the left, overran the left edge of the runway, crossed over the Washington Luís Avenue, and collided with a building, and with a fuel service station. All 187 persons on board and 12 people on the ground were killed in the accident.

Trans World Airlines (TWA) Flight 800

On July 17, 1996, about 2031 eastern daylight time, Trans World Airlines, Inc. (TWA) flight 800, a Boeing 747, crashed in the Atlantic Ocean near East Moriches, New York. TWA flight 800 was a scheduled international passenger flight from John F. Kennedy International Airport (JFK), New York, New York, to Charles DeGaulle International Airport, Paris, France. All 230 people on board were killed, and the airplane was destroyed. The weather was good. The National Transportation Safety Board determines that the probable cause of the accident was an explosion of the center wing fuel tank.

Turkish Airlines Flight TK1951

On 25 February 2009 a Boeing 737-800, flight TK1951, operated by Turkish Airlines was flying from Istanbul in Turkey to Amsterdam Schiphol Airport. There were 135 people on board. During the approach to the runway at Schiphol airport, the aircraft crashed about 1.5 kilometers from the threshold of the runway. This accident cost the lives of four crew members, and five passengers, 120 people sustained injuries. The crash was caused by a malfunctioning radio altimeter and a failure to implement the stall recovery procedure correctly.

United Airlines Flight 232

On July 19, 1989, a United Airlines' DC-10-10, on its way from Denver to Chicago, experienced a catastrophic failure of the No. 2 tail-mounted engine during cruise flight. The heroic pilots did all they could to bring the flight to a good end. But, notwithstanding all the attempts, the airplane subsequently crashed during an attempted landing at Sioux Gateway Airport, Iowa. Of the 296 people on board 111 were killed.

United Airlines Flight 585
This amended report explains the accident involving United Airlines flight 585, a Boeing 737-200, on its way from Denver to Colorado Springs, which crashed on March 3, 1991 near Colorado Springs Municipal Airport. Only after the crash of USAir 427 in 1994 and a similar incident with Eastwind 517 in 1996 the NTSB was able to pinpoint the cause of this crash: jammed rudder. The Boeing 737 has a history of rudder system-related anomalies, this finally solved the mystery of sudden jamming of the rudders of this aircraft.

United Airlines Flight 811
On February 24, 1989, United Airlines flight 811, a Boeing 747-122, lost a cargo door while climbing between 22,000 and 23,000 feet after taking off from Honolulu, Hawaii, with 355 persons aboard. Nine of the passengers were ejected from the airplane and lost at sea. The probable cause of this accident was a faulty switch in the door control system. Contributing to the cause of the accident was a deficiency in the design of the locking mechanisms, and a lack of timely corrective actions by Boeing and the FAA following an earlier incident.

Union des Transport Aériens de Guinée Flight GIH 141
On 25 December 2003, Union des Transport Aériens de Guinée Flight GIH 141, a Boeing 727-223, on a flight from Conakry (Guinea) to Kufra (Libya), Beirut (Lebanon) and Dubai (United Arab Emirates) stopped over at Cotonou, Republic of Benin. During takeoff the airplane, overloaded in an anarchic manner, was not able to climb properly and struck an airport building on the extended runway centerline, and crashed onto the beach, killing 151 of the 163 people on board. The crew was unknown with the forward center of gravity of the aircraft.

USAir Flight 427
The Boeing 737 has a history of rudder system-related anomalies, including numerous instances of jamming. During the course of the four and a half year investigation of the crash of USAir Flight 427 near Aliquippa, Pennsylvania, killing 132 people, the NTSB discovered that the PCU's dual servo valve could jam as well as deflect the rudder in the opposite direction of the pilots' input, due to thermal shock, caused when cold PCUs are injected with hot hydraulic fluid. This finally solved the mystery of sudden jamming of the rudders of this aircraft.

US Airways Flight 1549
On January 15, 2009, about 1527 eastern standard time, US Airways flight 1549, an Airbus Industrie A320-214, experienced an almost complete loss of thrust in both engines after encountering a flock of birds and was subsequently ditched on the Hudson River about 8.5 miles from LaGuardia Airport (LGA), New York City, New York. The flight was en route to Charlotte Douglas International Airport, Charlotte, North Carolina, and had departed LGA about 2 minutes

before the in-flight event occurred. The 150 passengers and 5 crewmembers evacuated the airplane via the forward and over wing exits.

ValuJet Airlines flight 592

On May 11, 1996, at 1413:42 eastern daylight time, a Douglas DC-9-32 crashed into the Everglades 10 minutes after takeoff from Miami International Airport, Miami, Florida. The airplane was being operated by ValuJet Airlines, Inc., as flight 592 and was on its way to Atlanta, Georgia. Both pilots, the three flight attendants, and all 105 passengers were killed. The NTSB determined that the cause of the accident, was a fire in the airplane's cargo compartment, initiated by the actuation of an oxygen generator being improperly carried as cargo.

0-0-0

www.ingramcontent.com/pod-product-compliance
Lightning Source LLC
Chambersburg PA
CBHW070617220526
45466CB00001B/27